SOLDIER TRAINING
PUBLICATION
No. 11-25C13-SM-TG

STP 11-25C13-SM-TG

HEADQUARTERS
DEPARTMENT OF THE ARMY
Washington, DC, 16 February 2005

SOLDIER'S MANUAL AND TRAINER'S GUIDE

MOS 25C

Radio Operator-Maintainer

Skill Levels 1, 2, and 3

TABLE OF CONTENTS

PAGE

Preface	vi
Chapter 1. Introduction	1-1
Chapter 2. Trainer's Guide	2-1
Chapter 3. MOS/Skill Level Tasks	3-1

Skill Level 1

Subject Area 1: RADIO PROCEDURES/MESSAGE HANDLING

113-571-1003	Communicate in a Radio Net	3-1
113-572-1002	Send a Message in 16-Line Format	3-3
113-573-6001	Recognize Electronic Attack (EA) and Implement Electronic Protection (EP)	3-4
113-573-7017	Submit Interference Message Report	3-7
113-572-1003	Operate VIASAT Software	3-10

Subject Area 2: ANTENNAS

113-596-1052	Construct Vertical Half-Rhombic Antenna	3-12
113-596-1056	Construct a Long-Wire Antenna	3-14
113-596-1068	Install Antenna Group OE-254/GRC (Team Method)	3-15
113-596-1070	Construct a Doublet Antenna	3-16
113-596-1071	Construct a Near Vertical Incidence Skywave (NVIS) Antenna AS-2259/GRA	3-17

DISTRIBUTION RESTRICTION: Approved for public release; distribution is unlimited.

*This publication supersedes STP 11-31C-13-SM-TG, 29 August 1997.

Subject Area 3: DATA COMMUNICATIONS EQUIPMENT

113-598-1023	Install Terminal Communications AN/UGC-74B(V)3	3-19
113-598-2010	Operate Terminal Communications AN/UGC-74B(V)3	3-20
113-598-2024	Perform Unit Level Maintenance on Communications Terminal AN/UGC-74B(V)3	3-22
113-609-2214	Operate Digital Message Data Generator, KL-43, or Similar Device	3-23
113-620-1044	Install Communications Central AN/TSC-128	3-26
113-620-2049	Operate Communications Central AN/TSC-128	3-28
113-620-3064	Perform Unit Level Maintenance on Communications Central AN/TSC-128	3-30

Subject Area 4: GENERATORS

113-601-1039	Install Generator Set AN/MJQ-35 5 KW	3-32
113-601-1040	Install Generator Set AN/MJQ-37 10 KW	3-34
113-601-1041	Install Generator Set AN/MJQ-42 3 KW	3-36
113-601-2054	Operate Generator Set AN/MJQ-35 5 KW	3-38
113-601-2055	Perform Operator's Troubleshooting Procedures on Generator Set AN/MJQ-35 5 KW	3-40
113-601-2057	Operate Generator Set AN/MJQ-37 10 KW	3-41
113-601-2058	Perform Operator's Troubleshooting Procedures and Generator Set AN/MJQ-37 10 KW	3-43
113-601-2059	Operate Generator Set AN/MJQ-42 3 KW	3-44
113-601-2060	Perform Operator's Troubleshooting Procedures on Generator Set AN/MJQ-42 3 KW	3-46
113-601-3016	Perform Unit Level Maintenance on Generator Sets 3 KW - 10 KW	3-47
113-601-3030	Perform PMCS on Generator Set AN/MJQ-35 5 KW	3-48
113-601-3031	Perform PMCS on Generator Set AN/MJQ-37 10 KW	3-50
113-601-3032	Perform PMCS on Generator Set AN/MJQ-42 3 KW	3-51

Subject Area 5: HIGH FREQUENCY (HF) RADIOS

113-620-0108	Perform Unit Level Troubleshooting on AN/PRC-150 Radio Set	3-52
113-620-1001	Install Radio Set AN/GRC-106(*)	3-54
113-620-1028	Install Radio Set AN/GRC-193A or Similar Radio Sets	3-56
113-620-2001	Operate Radio Set AN/GRC-106(*)	3-58
113-620-2002	Perform Operator Troubleshooting Procedures on Radio Set AN/GRC-106(*)	3-60
113-620-2028	Operate Radio Set AN/GRC-193A or Similar Radio Sets	3-61
113-620-2051	Operate Radio Set AN/PRC-150 Radio Set in Singal-Channel Mode	3-63
113-620-2052	Operate AN/PRC-150() in Automatic Link Establishment(ALE) Mode	3-65
113-620-3001	Perform PMCS on Radio Set AN/GRC-106(*)	3-67
113-620-3052	Perform Unit Level Maintenance for AN/GRC-193A or Similar Radio Sets	3-69

Subject Area 6: COMBAT NET RADIOS

113-587-1064	Prepare SINCGARS (Manpack) for Operation	3-71
113-587-2070	Operate SINCGARS Single-Channel (SC)	3-72
113-587-2071	Operate SINCGARS Frequency Hopping (FH) (Net Members)	3-74
113-587-2072	Operate SINCGARS Frequency Hopping (FH) Net Control Station (NCS)	3-76
113-587-2077	Operate SINCGARS Remote Control Unit (SRCU)	3-77

113-587-0070	Troubleshoot Secure Single-Channel Ground and Airborne Radio Systems (SINCGARS) ICOM With or Without the AN/VIC-1 or AN/VIC-3	3-78
113-589-0045	Troubleshoot Single-Channel Demand Assigned Multiple Access (DAMA) Tactical Satellite Terminal AN/PSC-5	3-80
113-589-2028	Operate Secure Single-Channel Tactical Satellite Radio Set in SATCOM Mode	3-82
113-589-2029	Operate Secure Single-Channel Tactical Satellite Radio Set in Demand Assigned Multiple Access (DAMA) Mode	3-85
113-589-3045	Perform Scheduled Unit Level Maintenance (ULM) on Single-Channel Demand Assigned Multiple Access (DAMA) Tactical Satellite Terminal AN/PSC-5	3-90
113-609-2053	Operate Automated Net Control Device (ANCD) AN/CYZ-10	3-92
113-609-2052	Perform Net Control Station (NCS) Duties Using Automated Net Control Device (ANCD) AN/CYZ-10	3-94
113-610-2044	Navigate Using the AN/PSN-11	3-96

Subject Area 7: AUTOMATION OPERATIONS

113-580-1032	Configure a Desktop IBM or Compatible Microcomputer for Operation	3-98
113-580-1033	Install Network Hardware/Software in a Desktop IBM or Compatible Microcomputer	3-100
113-580-1035	Install a Tactical Local Area Network (LAN)	3-102

Subject Area 8: REMOTE COMMUNICATIONS

113-622-1006	Install Radio Set Control Group AN/GRA-39(*)	3-104
113-622-2004	Operate Radio Set Control Group AN/GRA-39(*)	3-106
113-622-3001	Perform Unit Level Maintenance on Radio Set Control Group AN/GRA-39(*)	3-108

Subject Area 15: ABCS CORE

113-580-1055	Install Force XXI Battle Command Brigade and Below (FBCB2)	3-110
171-147-0001	PREPARE/SEND COMBAT MESSAGES USING FBCB2 VERSION 3.4	3-112
171-147-0002	PERFORM STARTUP PROCEDURES FOR FORCE XXI BATTLE COMMAND BRIGADE AND BELOW (FBCB2) VERSION 3.4	3-118
171-147-0005	APPLY MESSAGE ADDRESSING FEATURES IN FBCB2 VERSION 3.4	3-122
171-147-0006	PERFORM MESSAGE MANAGEMENT USING FBCB2 VERSION 3.4	3-126
171-147-0007	PREPARE/SEND OVERLAYS USING FBCB2 VERSION 3.4	3-128
171-147-0008	PREPARE/SEND REPORTS USING FBCB2 VERSION 3.4	3-134
171-147-0009	PREPARE/SEND FIRE/ALERT MESSAGES USING FBCB2 VERSION 3.4	3-138
171-147-0010	PREPARE/SEND ORDER/REQUEST MESSAGES USING FBCB2 VERSION 3.4	3-142
171-147-0011	PERFORM BEFORE-OPERATIONS PREVENTIVE MAINTENANCE CHECKS AND SERVICES ON FBCB2 VERSION 3.4	3-147
171-147-0012	PERFORM SHUT-DOWN PROCEDURES FOR FBCB2 VERSION 3.4	3-149
171-147-0013	PERFORM DURING-OPERATIONS PREVENTIVE MAINTENANCE CHECKS AND SERVICES ON FBCB2 VERSION 3.4	3-151
171-147-0014	PERFORM AFTER-OPERATIONS PREVENTIVE MAINTENANCE CHECKS AND SERVICES ON FBCB2 VERSION 3.4	3-152
171-147-0015	PREPARE/SEND A LOGISTICAL STATUS REPORT USING FBCB2 VERSION 3.4	3-153
171-147-0017	EMPLOY MAP FUNCTIONS USING FBCB2 VERSION 3.4	3-156
171-147-0019	EMPLOY FIPR FUNCTIONS USING FBCB2 VERSION 3.4	3-161
171-147-0020	EMPLOY STATUS FUNCTIONS USING FBCB2 VERSION 3.4	3-165
171-147-0021	EMPLOY ADMIN FUNCTIONS USING FBCB2 VERSION 3.4	3-168

171-147-0022 EMPLOY APPS FUNCTIONS USING FBCB2 VERSION 3.4 3-172
171-147-0023 EMPLOY NAV FUNCTIONS USING FBCB2 VERSION 3.4 3-177
171-147-0024 EMPLOY QUICK SEND FUNCTIONS USING FBCB2 VERSION 3.4 3-184
171-147-0025 EMPLOY FILTERS FUNCTIONS USING FBCB2 VERSION 3.4 3-186

Skill Level 2
Subject Area 9: COMBAT COMMUNICATIONS PLANNING

113-587-7133 Direct Implementation of an FM Voice Data Communications Network 3-189
113-589-7120 Direct Implementation of a Single-Channel Tactical Satellite
 Communications Network 3-191
113-611-1001 Select Team Radio Site 3-193
113-611-6112 Prepare Input to Signal Annex (Paragraph 5) of the OPORD 3-195
113-620-7089 Direct Implementation of a High Frequency (HF) Communications Network 3-197

Subject Area 12: CONSTRUCT ANTENNAS

113-596-1098 Construct Field Expedient Antennas 3-199

Subject Area 13: AUTOMATION OPERATIONS

113-580-0056 Troubleshoot Local Area Network (LAN) 3-201

Skill Level 3
Subject Area 9: COMBAT COMMUNICATIONS PLANNING

113-611-6002 Plan FM Voice and Data Communications Net 3-202
113-611-6004 Plan a Single-Channel Tactical Satellite Communications Network 3-204
113-611-6005 Plan an HF Communications Network 3-206

Subject Area 10: RADIO NET OPERATION

113-571-7002 Inspect Station/Net Operation and Duties 3-208
113-573-1004 Conduct Communications Security Inspections 3-210
113-596-7081 Inspect Installation of Antennas 3-212
113-620-7088 Inspect Installed Operational Radio Sets 3-213

Subject Area 11: GENERATORS AND EQUIPMENT PREVENTATIVE MAINTENANCE CHECKS AND SERVICES

113-601-7055 Inspect Installed Operational Generator Sets 3-215
113-617-7119 Inspect the Preventive Maintenance Checks and Services (PMCS) 3-216

Chapter 4. Duty Position Tasks 4-1

Subject Area 14: ENHANCED POSITION LOCATION REPORTING SYSTEM

113-601-2056 Operate Generator Set MEP-803A 4-1
113-630-1002 Prepare the EPLRS Net Control Station for Movement 4-3
113-630-1005 Install the EPLRS Net Control Station 4-5
113-630-1006 Deploy the EPLRS Gateway Radio Set 4-7
113-630-1007 Deploy the Grid Reference Radio Set 4-10
113-630-2001 Operate the EPLRS Net Control Station 4-12
113-630-2027 Perform ECCM Procedures 4-14
113-630-2031 Perform Playback Mode Operational Procedures 4-15
113-630-2044 Operate the EPLRS Radio Set 4-16
113-630-2045 Perform Multiple Net Control Station (NCS) Community Operational
 Procedures 4-18

    113-630-3011 Perform Preventive Maintenance Checks and Services (PMCS) on the
EPLRS Net Control Station... 4-20
    113-630-3014 Troubleshoot the EPLRS Net Control Station 4-22

Appendix A - DA Form 5164-R (Hands-On Evaluation)... A-1

Appendix B - DA Form 5165-R (Field Expedient Book)... B-1

Glossary Glossary-1

References... References-1

PREFACE

This publication is for skill levels (SLs) 1, 2, and 3 Soldiers holding military occupational specialty (MOS) 25C and for trainers and first-line supervisors. It contains standardized training objectives, in the form of task summaries, to train and evaluate Soldiers on critical tasks that support unit missions during wartime. Trainers and first-line supervisor should ensure Soldiers holding MOS 25C SLs 1, 2, and 3 have access to this publication. It should be made available in the Soldier's work area, unit learning center, and unit libraries.

This manual applies to both Active and Reserve Component Soldiers.

The proponent for this publication is the Signal School. Send comments and recommendations on DA Form 2028 (Recommended Changes to Publications and Blank Forms) directly to Commander, US Army Signal Center and Fort Gordon, ATTN: ATZH-DTM, Fort Gordon, Georgia 30905-5735.

This manual is available on the General Dennis J. Reimer Training and Doctrine Digital Library for viewing and downloading. The WWW address is http://www.adtdl.army.mil/.

Unless this manual states otherwise, masculine pronouns do not refer exclusively to men.

NOTE: Information contained in this publication is subject to change as new equipment is added to the Army inventory and revisions in policy and doctrine are made.

STP 11-25C13-SM-TG

CHAPTER 1

Introduction

1-1. GENERAL. The Soldier training publication (STP) identifies the individual military occupational specialty (MOS) and training requirements for Soldiers in various specialties. Another source of STP task data is the General Dennis J. Reimer Training and Doctrine Digital Library at http://www.adtdl.army.mil/. Commanders, trainers, and Soldiers should use the STP to plan, conduct, and evaluate individual training in units. The STP is the primary MOS reference to support the self-development and training of every Soldier in the unit. It is used with the Soldier's Manual of Common Tasks, Army training and evaluation programs (ARTEPs), and FM 7-0, *Training the Force*, to establish effective training plans and programs that integrate Soldier, leader, and collective tasks. This chapter explains how to use the STP in establishing an effective individual training program. It includes doctrinal principles and implications outlined in FM 7-0. Based on these guidelines, commanders and unit trainers must tailor the information to meet the requirements for their specific unit.

1-2. TRAINING REQUIREMENT. Every Soldier, noncommissioned officer (NCO), warrant officer, and officer has one primary mission—to be trained and ready to fight and win our nation's wars. Success in battle does not happen by accident; it is a direct result of tough, realistic, and challenging training.

    a. Operational Environment.

    (1) Commanders and leaders at all levels must conduct training with respect to a wide variety of operational missions across the full spectrum of operations; these operations may include combined arms, joint, multinational, and interagency considerations, and span the entire breadth of terrain and environmental possibilities. Commanders must strive to set the daily training conditions as closely as possible to those expected for actual operations.

    (2) The operational missions of the Army include not only war, but also military operations other than war (MOOTW). Operations may be conducted as major combat operations, a small-scale contingency, or a peacetime military engagement. Offensive and defensive operations normally dominate military operations in war along with some small-scale contingencies. Stability operations and support operations dominate in MOOTW. Commanders at all echelons may combine different types of operations simultaneously and sequentially to accomplish missions in war and MOOTW. These missions require training since future conflict will likely involve a mix of combat and MOOTW, often concurrently. The range of possible missions complicates training. Army forces cannot train for every possible mission; they train for war and prepare for specific missions as time and circumstances permit.

    (3) Our forces today use a train-alert-deploy sequence. We cannot count on the time or opportunity to correct or make up training deficiencies after deployment. Maintaining forces that are ready now, places increased emphasis on training and the priority of training. This concept is a key link between operational and training doctrine.

    (4) Units train to be ready for war based on the requirements of a precise and specific mission; in the process they develop a foundation of combat skills that can be refined based on the requirements of the assigned mission. Upon alert, commanders assess and refine from this foundation of skills. In the train-alert-deploy process, commanders use whatever time the alert cycle provides to continue refinement of mission-focused training. Training continues during time available between alert notification and deployment, between deployment and employment, and even during employment as units adapt to the specific battlefield environment and assimilate combat replacements.

    b. How the Army Trains the Army.

    (1) Training is a team effort and the entire Army—Department of the Army (DA), major commands (MACOMs), the institutional training base, units, the combat training centers (CTCs), each individual

Soldier and the civilian workforce—has a role that contributes to force readiness. DA and MACOMs are responsible for resourcing the Army to train. The Institutional Army, including schools, training centers, and NCO academies, for example, train Soldiers and leaders to take their place in units in the Army by teaching the doctrine and tactics, techniques, and procedures (TTP). Units, leaders, and individuals train to standard on their assigned critical individual tasks. The unit trains first as an organic unit and then as an integrated component of a team. Before the unit can be trained to function as a team, each Soldier must be trained to perform their individual supporting tasks to standard. Operational deployments and major training opportunities, such as major training exercises, CTCs, and ARTEPs provide rigorous, realistic, and stressful training and operational experience under actual or simulated combat and operational conditions to enhance unit readiness and produce bold, innovative leaders. The result of this Armywide team effort is a training and leader development system that is unrivaled in the world. Effective training produces the force—Soldiers, leaders, and units—that can successfully execute any assigned mission.

(2) The Army Training and Leader Development Model (Figure 1-1) centers on developing trained and ready units led by competent and confident leaders. The model depicts an important dynamic that creates a lifelong learning process. The three core domains that shape the critical learning experiences throughout a Soldiers and leaders time span are the operational, institutional, and self-development domains. Together, these domains interact using feedback and assessment from various sources and methods to maximize warfighting readiness. Each domain has specific, measurable actions that must occur to develop our leaders.

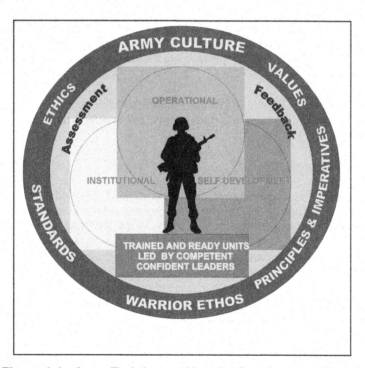

Figure 1-1. Army Training and Leader Development Model

(3) The operational domain includes home station training, CTC rotations, and joint training exercises and deployments that satisfy national objectives. Each of these actions provides foundational experiences for Soldier, leader, and unit development. The institutional domain focuses on educating and training Soldiers and leaders on the key knowledge, skills, and attributes required to operate in any environment. It includes individual, unit and joint schools, and advanced education. The self-development domain, both structured and informal, focuses on taking those actions necessary to reduce or eliminate the gap between operational and institutional experiences.

(4) Throughout this lifelong learning and experience process, there is formal and informal assessment and feedback of performance to prepare leaders and Soldiers for their next level of responsibility. Assessment is the method used to determine the proficiency and potential of leaders against a known standard. Feedback must be clear, formative guidance directly related to the outcome of training events measured against standards.

    c. Leader Training and Leader Development.

(1) Competent and confident leaders are a prerequisite to the successful training of units. It is important to understand that leader training and leader development are integral parts of unit readiness. Leaders are inherently Soldiers first and should be technically and tactically proficient in basic Soldier skills. They are also adaptive, capable of sensing their environment, adjusting the plan when appropriate, and properly applying the proficiency acquired through training.

(2) Leader training is an expansion of these skills that qualifies them to lead other Soldiers. As such, doctrine and principles of training require the same level of attention of senior commanders. Leader training occurs in the Institutional Army, the unit, the CTCs, and through self-development. Leader training is just one portion of leader development.

(3) Leader development is the deliberate, continuous, sequential, and progressive process, grounded in Army values, that grows Soldiers and civilians into competent and confident leaders capable of decisive action. Leader development is achieved through the lifelong synthesis of the knowledge, skills, and experiences gained through institutional training and education, organizational training, operational experience, and self-development. Commanders play the key roll in leader development that ideally produces tactically and technically competent, confident, and adaptive leaders who act with boldness and initiative in dynamic, complex situations to execute mission-type orders achieving the commander's intent.

    d. Training Responsibility. Soldier and leader training and development continue in the unit. Using the institutional foundation, training in organizations and units focuses and hones individual and team skills and knowledge.

(1) **Commander Responsibility**.

(a) The unit commander is responsible for the wartime readiness of all elements in the formation. The commander is, therefore, the primary trainer of the organization and is responsible for ensuring that all training is conducted in accordance with the STP to the Army standard.

(b) Commanders ensure STP standards are met during all training. If a Soldier fails to meet established standards for identified MOS tasks, the Soldier must retrain until the tasks are performed to standard. Training to standard on MOS tasks is more important than completion of a unit-training event such as an ARTEP. The objective is to focus on sustaining MOS proficiency—this is the critical factor commanders must adhere to when training individual Soldiers units.

(2) **NCO Responsibility**.

(a) A great strength of the US Army is its professional NCO Corps who takes pride in being responsible for the individual training of Soldiers, crews, and small teams. The NCO support channel parallels and complements the chain of command. It is a channel of communication and supervision from the Command Sergeant Major (CSM) to the First Sergeants (1SGs) and then to other NCOs and enlisted personnel. NCOs train Soldiers to the nonnegotiable standards published in STPs. Commanders delegate authority to NCOs in the support channel as the primary trainers of individual, crew, and small team training. Commanders hold NCOs responsible for conducting standards-based, performance-oriented, battle-focused training and providing feedback on individual, crew, and team proficiency. Commanders define responsibilities and authority of their NCOs to their staffs and subordinates.

(b) NCOs continue the Soldierization process of newly assigned enlisted Soldiers, and begin their professional development. NCOs are responsible for conducting standards-based, performance-oriented, battle-focused training. They identify specific individual, crew, and small team tasks that support the unit's collective mission essential tasks; plan, prepare, rehearse, and execute training; and evaluate training and conduct after action reviews (AARs) to provide feedback to the commander on individual, crew, and small team proficiency. Senior NCOs coach junior NCOs to master a wide range of individual tasks.

(3) **Soldier Responsibility**. Each Soldier is responsible for performing individual tasks identified by the first-line supervisor based on the unit's mission essential task list (METL). Soldiers must perform tasks to the standards included in the task summary. If Soldiers have questions about tasks or which tasks in this manual they must perform, they are responsible for asking their first-line supervisor for clarification, assistance, and guidance. First-line supervisors know how to perform each task or can direct Soldiers to appropriate training materials, including current field manuals, technical manuals, and Army regulations. Soldiers are responsible for using these materials to maintain performance. They are also responsible for maintaining standard performance levels of all Soldiers' Manual of Common Tasks at their current skill level and below. Periodically, Soldiers should ask their supervisor or another Soldier to check their performance to ensure that they can perform the tasks.

1-3. **BATTLE-FOCUSED TRAINING**. Battle focus is a concept used to derive peacetime training requirements from assigned and anticipated missions. The priority of training in units is to train to standard on the wartime mission. Battle focus guides the planning, preparation, execution, and assessment of each organization's training program to ensure its members train as they are going to fight. Battle focus is critical throughout the entire training process and is used by commanders to allocate resources for training based on wartime and operational mission requirements. Battle focus enables commanders and staffs at all echelons to structure a training program that copes with non-mission-related requirements while focusing on mission essential training activities. It is recognized that a unit cannot attain proficiency to standard on every task whether due to time or other resource constraints. However, unit commanders can achieve a successful training program by consciously focusing on a reduced number of METL tasks that are essential to mission accomplishment.

a. Linkage Between METL and STP. A critical aspect of the battle focus concept is to understand the responsibility for and the linkage between the collective mission essential tasks and the individual tasks that support them. For example, the commander and the CSM/1SG must jointly coordinate the collective mission essential tasks and supporting individual tasks on which the unit will concentrate its efforts during a given period. This task hierarchy is provided in the task database at the Reimer Digital Library. The CSM/1SG must select the specific individual tasks that support each collective task to be trained. Although NCOs have the primary role in training and sustaining individual Soldier skills, officers at every echelon remain responsible for training to established standards during both individual and collective training. Battle focus is applied to all missions across the full spectrum of operations.

b. Relationship of STPs to Battle-Focused Training. The two key components of any STP are the Soldier's manual (SM) and trainer's guide (TG). Each gives leaders important information to help implement the battle-focused training process. The training guide relates Soldier and leader tasks in the MOS and skill level to duty positions and equipment. It states where the task is trained, how often training should occur to sustain proficiency, and who in the unit should be trained. As leaders assess and plan training, they should rely on the training guide to help identify training needs.

(1) Leaders conduct and evaluate training based on Armywide training objectives and on the task standards published in the Soldier's manual task summaries or in the Reimer Digital Library. The task summaries ensure that trainers in every unit and location define task standards the same way and trainers evaluate all Soldiers to the same standards.

(2) Figure 1-2 shows how battle-focused training relates to the training guide and Soldier's manual. The left column shows the steps involved in training Soldiers and the right column shows how the STP supports each of these steps.

BATTLE-FOCUS PROCESS	STP SUPPORT PROCESS
Select supporting Soldier tasks	Use TG to relate tasks to METL
Conduct training assessment	Use TG to define what Soldier tasks to assess
Determine training objectives	Use TG to set objectives
Determine strategy; plan for training	Use TG to relate Soldier tasks to strategy
Conduct pre-execution checks	Use SM task summary as source for task performance
Execute training; conduct after action review	Use SM task summary as source for task performance
Evaluate training against established standards	Use SM task summary as standard for evaluation

Figure 1-2. Relationship of Battle-Focused Training and STP

1-4. **TASK SUMMARY FORMAT**. Task summaries outline the wartime performance requirements of each critical task in the SM. They provide the Soldier and the trainer with the information necessary to prepare, conduct, and evaluate critical task training. As a minimum, task summaries include information the Soldier must know and the skills that he must perform to standards for each task. The format of the task summaries included in this SM is as follows:

 a. Task Number. A 10-digit number identifies each task or skill. This task number, along with the task title, must be included in any correspondence pertaining to the task.

 b. Task Title. The task title identifies the action to be performed.

 c. Conditions. The task conditions identify all the equipment, tools, references, job aids, and supporting personnel that the Soldier needs to use to perform the task in wartime. This section identifies any environmental conditions that can alter task performance, such as visibility, temperature, or wind. This section also identifies any specific cues or events that trigger task performance, such as a chemical attack or identification of a threat vehicle.

 d. Standards. The task standards describe how well and to what level the task must be performed under wartime conditions. Standards are typically described in terms of accuracy, completeness, and speed.

 e. Training and Evaluation. The training evaluation section identifies specific actions, known as performance steps, that the Soldier must do to successfully complete the task. These actions are in the evaluation guide section of the task summary and are listed in a GO/NO GO format for easy evaluation. For some tasks, the training and evaluation section may also include detailed training information in a training information outline and an evaluation preparation section. The evaluation preparation section indicates necessary modifications to task performance in order to train and evaluate a task that cannot be trained to the wartime conditions. It may also include special training and evaluation preparation instructions to accommodate these modifications, and any instructions that should be given to the Soldier before evaluation.

 f. References. This section identifies references that provide more detailed and thorough explanations of task performance requirements than those given in the task summary description.

g. Warnings. Warnings alert users to the possibility of immediate personal injury or damage to equipment.

h. Notes. Notes provide a supportive explanation or hint that relates to the performance standards.

1-5. TRAINING EXECUTION. All good training, regardless of the specific collective, leader, and individual tasks being executed, must comply with certain common requirements. These include adequate preparation, effective presentation and practice, and thorough evaluation. The execution of training includes preparation for training, conduct of training, and recovery from training.

a. Preparation for Training. Formal near-term planning for training culminates with the publication of the unit-training schedule. Informal planning, detailed coordination, and preparation for executing the training continue until the training is performed. Commanders and other trainers use training meetings to assign responsibility for preparation of all scheduled training. Preparation for training includes selecting tasks to be trained, planning the conduct of the training, training the trainers, reconnaissance of the site, issuing the training execution plan, and conducting rehearsals and pre-execution checks. Pre-execution checks are preliminary actions commanders and trainers use to identify responsibility for these and other training support tasks. They are used to monitor preparation activities and to follow up to ensure planned training is conducted to standard. Pre-execution checks are a critical portion of any training meeting. During preparation for training, battalion and company commanders identify and eliminate potential training distracters that develop within their own organizations. They also stress personnel accountability to ensure maximum attendance at training.

(1) Subordinate leaders, as a result of the bottom-up feed from internal training meetings, identify and select the individual tasks necessary to support the identified training objectives. Commanders develop the tentative plan to include requirements for preparatory training, concurrent training, and training resources. At a minimum, the training plan should include confirmation of training areas and locations, training ammunition allocations, training simulations and simulators availability, transportation requirements, Soldier support items, a risk management analysis, assignment of responsibility for the training, designation of trainers responsible for approved training, and final coordination. The time and other necessary resources for retraining must also be an integral part of the original training plan.

(2) Leaders, trainers, and evaluators are identified, trained to standard, and rehearsed prior to the conduct of the training. Leaders and trainers are coached on how to train, given time to prepare, and rehearsed so that training will be challenging and doctrinally correct. Commanders ensure that trainers and evaluators are not only tactically and technically competent on their training tasks, but also understand how the training relates to the organization's METL. Properly prepared trainers, evaluators, and leaders project confidence and enthusiasm to those being trained. Trainer and leader training is a critical event in the preparation phase of training. These individuals must demonstrate proficiency on the selected tasks prior to the conduct of training.

(3) Commanders, with their subordinate leaders and trainers, conduct site reconnaissance, identify additional training support requirements, and refine and issue the training execution plan. The training plan should identify all those elements necessary to ensure the conduct of training to standard. Rehearsals are essential to the execution of good training. Realistic, standards-based, performance-oriented training requires rehearsals for trainers, support personnel, and evaluators. Preparing for training in Reserve Component (RC) organizations can require complex pre-execution checks. RC trainers must often conduct detailed coordination to obtain equipment, training support system products, and ammunition from distant locations. In addition, RC pre-execution checks may be required to coordinate Active Component (AC) assistance from the numbered Armies in the continental United States (CONUS), training support divisions, and directed training affiliations.

b. Conduct of Training. Ideally, training is executed using the crawl-walk-run approach. This allows and promotes an objective, standards-based approach to training. Training starts at the basic level. Crawl events are relatively simple to conduct and require minimum support from the unit. After the

STP 11-25C13-SM-TG

crawl stage, training becomes incrementally more difficult, requiring more resources from the unit and home station, and increasing the level of realism. At the run stage, the level of difficulty for the training event intensifies. Run stage training requires optimum resources and ideally approaches the level of realism expected in combat. Progression from the walk to the run stage for a particular task may occur during a one-day training exercise or may require a succession of training periods over time. Achievement of the Army standard determines progression between stages.

(1) In crawl-walk-run training, the tasks and the standards remain the same; however, the conditions under which they are trained change. Commanders may change the conditions, for example, by increasing the difficulty of the conditions under which the task is being performed, increasing the tempo of the task training, increasing the number of tasks being trained, or by increasing the number of personnel involved in the training. Whichever approach is used, it is important that all leaders and Soldiers involved understand in which stage they are currently training and understand the Army standard.

(2) An AAR is immediately conducted and may result in the need for additional training. Any task that was not conducted to standard should be retrained. Retraining should be conducted at the earliest opportunity. Commanders should program time and other resources for retraining as an integral part of their training plan. Training is incomplete until the task is trained to standard. Soldiers will remember the standard enforced, not the one discussed.

c. Recovery from Training. The recovery process is an extension of training, and once completed, it signifies the end of the training event. At a minimum, recovery includes conduct of maintenance training, turn-in of training support items, and the conduct of AARs that review the overall effectiveness of the training just completed.

(1) Maintenance training is the conduct of post-operations preventive maintenance checks and services (PMCS), accountability of organizational and individual equipment, and final inspections. Class IV, Class V, Training Aids, Devices, Simulators, and Simulations (TADSS) and other support items are maintained, accounted for, and turned-in, and training sites and facilities are closed out.

(2) AARs conducted during recovery focus on collective, leader, and individual task performance, and on the planning, preparation, and conduct of the training just completed. Unit AARs focus on individual and collective task performance, and identify shortcomings and the training required to correct deficiencies. AARs with leaders focus on tactical judgment. These AARs contribute to leader learning and provide opportunities for leader development. AARs with trainers and evaluators provide additional opportunities for leader development.

1-6. **TRAINING ASSESSMENT.** Assessment is the commander's responsibility. It is the commander's judgment of the organization's ability to accomplish its wartime operational mission. Assessment is a continuous process that includes evaluating individual training, conducting an organizational assessment, and preparing a training assessment. The commander uses his experience, feedback from training evaluations, and other evaluations and reports to arrive at his assessment. Assessment is both the end and the beginning of the training management process. Training assessment is more than just training evaluation, and encompasses a wide variety of inputs. Assessments include such diverse systems as training, force integration, logistics, and personnel, and provide the link between the unit's performance and the Army standard. Evaluation of training is, however, a major component of assessment. Training evaluations provide the commander with feedback on the demonstrated training proficiency of Soldiers, leaders, battle staffs, and units. Commanders cannot personally observe all training in their organization and, therefore, gather feedback from their senior staff officers and NCOs.

a. Evaluation of Training. Training evaluations are a critical component of any training assessment. Evaluation measures the demonstrated ability of Soldiers, commanders, leaders, battle staffs, and units against the Army standard. Evaluation of training is integral to standards-based training and is the cornerstone of leader training and leader development. STPs describe standards that must be met for each Soldier task.

(1) All training must be evaluated to measure performance levels against the established Army standard. The evaluation can be as fundamental as an informal, internal evaluation performed by the leader conducting the training. Evaluation is conducted specifically to enable the individual undergoing the training to know whether the training standard has been achieved. Commanders must establish a climate that encourages candid and accurate feedback for the purpose of developing leaders and trained Soldiers.

(2) Evaluation of training is not a test; it is not used to find reasons to punish leaders and Soldiers. Evaluation tells Soldiers whether or not they achieved the Army standard and, therefore, assists them in determining the overall effectiveness of their training plans. Evaluation produces disciplined Soldiers, leaders, and units. Training without evaluation is a waste of time and resources.

(3) Leaders use evaluations as an opportunity to coach and mentor Soldiers. A key element in developing leaders is immediate, positive feedback that coaches and leads subordinate leaders to achieve the Army standard. This is a tested and proven path to develop competent, confident adaptive leaders.

b. Evaluators. Commanders must plan for formal evaluation and must ensure the evaluators are trained. These evaluators must also be trained as facilitators to conduct AARs that elicit maximum participation from those being trained. External evaluators will be certified in the tasks they are evaluating and normally will not be dual-hatted as a participant in the training being executed.

c. Role of Commanders and Leaders. Commanders ensure that evaluations take place at each echelon in the organization. Commanders use this feedback to teach, coach, and mentor their subordinates. They ensure that every training event is evaluated as part of training execution and that every trainer conducts evaluations. Commanders use evaluations to focus command attention by requiring evaluation of specific mission essential and battle tasks. They also take advantage of evaluation information to develop appropriate lessons learned for distribution throughout their commands.

d. After Action Review. The AAR, whether formal or informal, provides feedback for all training. It is a structured review process that allows participating Soldiers, leaders, and units to discover for themselves what happened during the training, why it happened, and how it can be done better. The AAR is a professional discussion that requires the active participation of those being trained. FM 7-0 provides detailed instructions for conducting an AAR and detailed guidance on coaching and critiquing during training.

1-7. NCO SELF-DEVELOPMENT AND THE SOLDIER'S MANUAL

a. Self-development is one of the key components of the leader development program. It is a planned progressive and sequential program followed by leaders to enhance and sustain their military competencies. It consists of individual study, research, professional reading, practice, and self-assessment. Under the self-development concept, the NCO, as an Army professional, has the responsibility to remain current in all phases of the MOS. The SM is the primary source for the NCO to use in maintaining MOS proficiency.

b. Another important resource for NCO self-development is the Army Correspondence Course Program (ACCP). Soldiers should refer to DA Pamphlet 350-59, *Army Correspondence Course Program Catalog*, for a list of courses and information on enrolling in this program, or contact ACCP Student Support at DSN 826-2127/3322, COML (757) 878-2127/3322, or E-mail at mailto:Sectiona@atsc.army.mil. Soldiers can also access the Army Correspondence Course Program online at http://www.atsc.army.mil/accp/aipdnew.asp.

c. General Dennis J. Reimer Training and Doctrine Digital Library is an additional resource for NCO self-development. This electronic library is the single repository of approved Army training and doctrine information. Soldiers can access the library online at http://www.adtdl.army.mil.

d. Unit learning centers are valuable resources for planning self-development programs. They can help access enlisted career maps, training support products, and extension training materials, such as field manuals (FMs) and technical manuals (TMs). It is the Soldier's responsibility to use these materials to maintain performance.

1-8. **TRAINING SUPPORT.** This manual includes the following appendixes and information that provide additional training support information.

a. Appendix A, DA Form 5164-R (Hands-On Evaluation). This appendix contains the instructions for using DA Form 5164-R and a sample completed form for NCOs to use during evaluation of Soldiers' manual tasks.

b. Appendix B, DA Form 5165-R (Field Expedient Squad Book). This appendix contains the instructions for using DA Form 5165-R and a sample completed form for NCOs to use during evaluation of Soldiers' manual tasks.

c. Glossary. The glossary is a single comprehensive list of acronyms, abbreviations, definitions, and letter symbols.

d. References. This section contains two lists of references, required and related, which support training of all tasks in this SM. Required references are listed in the conditions statement and are required for the Soldier to do the task. Related references are materials that provide more detailed information and a more thorough explanation of task performance.

1-9. **FEEDBACK.** Recommendations for improvement of this STP are requested. Feedback will help to ensure that this STP answers the training needs of units in the field.

This page intentionally left blank.

CHAPTER 2

Trainer's Guide

2-1. **GENERAL.** The MOS Training Plan (MTP) identifies the essential components of a unit-training plan for individual training. Units have different training needs and requirements based on differences in environment, location, equipment, dispersion, and similar factors. Therefore, the MTP should be used as a guide for conducting unit training and not a rigid standard. The MTP shows the relationship of an MOS SL between duty position and critical tasks. These critical tasks are grouped by task commonality into subject areas.

The MTP's Subject Area Codes list subject area numbers and titles used throughout the MTP. These subject areas are used to define the training requirements for each duty position within an MOS.

The Duty Position Training Requirements table identifies the total training requirement for each duty position within an MOS and provides a recommendation for cross training and train-up/merger training.

- **Duty Position column.** This column lists the duty positions of the MOS, by skill level, which have different training requirements.

- **Subject Area column.** This column lists, by numerical key, the subject areas a Soldier must be proficient in to perform in that duty position.

- **Cross Train column.** This column lists the recommended duty position for which Soldiers should be cross-trained.

- **Train-up/Merger column.** This column lists the corresponding duty position for the next higher skill level or MOSC the Soldier will merge into on promotion.

The Critical Task List table lists, by general subject areas, the critical tasks to be trained in an MOS and the type of training required (resident, integration, or sustainment).

- **Subject Area column.** This column lists the subject area number and title.

- **Task Number column.** This column lists the task numbers for all tasks included in the subject area.

- **Title column.** This column lists the task title for each task in the subject area.

- **Training Location column.** This column identifies the training location where the task is first trained to Soldier training publications standards. If the task is first trained to standard in the unit, the word "Unit" will be in this column. If the task is first trained to standard in the training base, it will identify, by brevity code (ANCOC, BNCOC, etc.), the resident course where the task was taught. Figure 2-1 contains a list of training locations and their corresponding brevity codes.

UNIT	Trained in the Unit
AIT	Advanced Individual Training
BNCOC	Basic NCO Course
ANCOC	Advanced NCO Course
ASI	Additional Skill Identifier

Figure 2-1. Training Locations

STP 11-25C13-SM-TG

- **Sustainment Training Frequency Column.** This column indicates the recommended frequency at which the tasks should be trained to ensure Soldiers maintain task proficiency. Figure 2-2 identifies the frequency codes used in this column.

```
BA - Biannually
AN - Annually
SA - Semiannually
QT - Quarterly
MO - Monthly
BW - Biweekly
WK - Weekly
```

Figure 2-2. Sustainment Training Frequency Codes

- **Sustainment Training Skill Level Column.** This column lists the skill levels of the MOS for which Soldiers must receive sustainment training to ensure they maintain proficiency to Soldier's manual standards.

2-2. **SUBJECT AREA CODES.**

Skill Level 1
 1 RADIO PROCEDURES/MESSAGE HANDLING
 2 ANTENNAS
 3 DATA COMMUNICATIONS EQUIPMENT
 4 GENERATORS
 5 HIGH FREQUENCY (HF) RADIOS
 6 COMBAT NET RADIOS
 7 AUTOMATION OPERATIONS
 8 REMOTE COMMUNICATIONS
 14 ENHANCED POSITION LOCATION REPORTING SYSTEM
 15 ABCS CORE

Skill Level 2
 9 COMBAT COMMUNICATIONS PLANNING
 12 CONSTRUCT ANTENNAS
 13 AUTOMATION OPERATIONS

Skill Level 3
 9 COMBAT COMMUNICATIONS PLANNING
 10 RADIO NET OPERATION
 11 GENERATORS AND EQUIPMENT PREVENTATIVE MAINTENANCE CHECKS AND SERVICES

STP 11-25C13-SM-TG

2-3. **CRITICAL TASKS LIST.**

**MOS TRAINING PLAN
25C13**

CRITICAL TASKS

Task Number	Title	Training Location	Sust Tng Freq	Sust Tng SL
Skill Level 1				
Subject Area 1. RADIO PROCEDURES/MESSAGE HANDLING				
113-571-1003	Communicate in a Radio Net	AIT	AN	1-3
113-572-1002	Send a Message in 16-Line Format	AIT	SA	1-3
113-573-6001	Recognize Electronic Attack (EA) and Implement Electronic Protection (EP)	AIT	QT	1-3
113-573-7017	Submit Interference Message Report	AIT	QT	1-3
113-572-1003	Operate VIASAT Software	AIT	AN	1-3
Subject Area 2. ANTENNAS				
113-596-1052	Construct Vertical Half-Rhombic Antenna	UNIT	AN	1-3
113-596-1056	Construct a Long-Wire Antenna	UNIT	AN	1-3
113-596-1068	Install Antenna Group OE-254/GRC (Team Method)	AIT	SA	1-3
113-596-1070	Construct a Doublet Antenna	AIT	SA	1-3
113-596-1071	Construct a Near Vertical Incidence Skywave (NVIS) Antenna AS-2259/GRA	AIT	SA	1-3
Subject Area 3. DATA COMMUNICATIONS EQUIPMENT				
113-598-1023	Install Terminal Communications AN/UGC-74B(V)3	UNIT	AN	1-3
113-598-2010	Operate Terminal Communications AN/UGC-74B(V)3	UNIT	AN	1-3
113-598-2024	Perform Unit Level Maintenance on Communications Terminal AN/UGC-74B(V)3	UNIT	AN	1-3
113-609-2214	Operate Digital Message Data Generator, KL-43, or Similar Device	AIT	AN	1-3
113-620-1044	Install Communications Central AN/TSC-128	UNIT	AN	1-3
113-620-2049	Operate Communications Central AN/TSC-128	UNIT	AN	1-3
113-620-3064	Perform Unit Level Maintenance on Communications Central AN/TSC-128	UNIT	AN	1-3
Subject Area 4. GENERATORS				
113-601-1039	Install Generator Set AN/MJQ-35 5 KW	AIT	AN	1-3
113-601-1040	Install Generator Set AN/MJQ-37 10 KW	UNIT	AN	1-3
113-601-1041	Install Generator Set AN/MJQ-42 3 KW	UNIT	AN	1-3
113-601-2054	Operate Generator Set AN/MJQ-35 5 KW	AIT	AN	1-3
113-601-2055	Perform Operator's Troubleshooting Procedures on Generator Set AN/MJQ-35 5 KW	AIT	AN	1-3
113-601-2057	Operate Generator Set AN/MJQ-37 10 KW	UNIT	AN	1-3
113-601-2058	Perform Operator's Troubleshooting Procedures and Generator Set AN/MJQ-37 10 KW	UNIT	AN	1-3
113-601-2059	Operate Generator Set AN/MJQ-42 3 KW	UNIT	AN	1-3

16 February 2005

CRITICAL TASKS

Task Number	Title	Training Location	Sust Tng Freq	Sust Tng SL
113-601-2060	Perform Operator's Troubleshooting Procedures on Generator Set AN/MJQ-42 3 KW	UNIT	AN	1-3
113-601-3016	Perform Unit Level Maintenance on Generator Sets 3 KW - 10 KW	UNIT	AN	1-3
113-601-3030	Perform PMCS on Generator Set AN/MJQ-35 5 KW	AIT	AN	1-3
113-601-3031	Perform PMCS on Generator Set AN/MJQ-37 10 KW	UNIT	AN	1-3
113-601-3032	Perform PMCS on Generator Set AN/MJQ-42 3 KW	UNIT	AN	1-3
Subject Area 5. HIGH FREQUENCY (HF) RADIOS				
113-620-0108	Perform Unit Level Troubleshooting on AN/PRC-150 Radio Set	AIT	AN	1-3
113-620-1001	Install Radio Set AN/GRC-106(*)	UNIT	AN	1-3
113-620-1028	Install Radio Set AN/GRC-193A or Similar Radio Sets	AIT	AN	1-3
113-620-2001	Operate Radio Set AN/GRC-106(*)	UNIT	AN	1-3
113-620-2002	Perform Operator Troubleshooting Procedures on Radio Set AN/GRC-106(*)	UNIT	AN	1-3
113-620-2028	Operate Radio Set AN/GRC-193A or Similar Radio Sets	AIT	AN	1-3
113-620-2051	Operate Radio Set AN/PRC-150 Radio Set in Single-Channel Mode	AIT	AN	1-3
113-620-2052	Operate AN/PRC-150( ) in Automatic Link Establishment(ALE) Mode	AIT	AN	1-3
113-620-3001	Perform PMCS on Radio Set AN/GRC-106(*)	UNIT	AN	1-3
113-620-3052	Perform Unit Level Maintenance for AN/GRC-193A or Similar Radio Sets	AIT	AN	1-3
Subject Area 6. COMBAT NET RADIOS				
113-587-1064	Prepare SINCGARS (Manpack) for Operation	AIT	AN	1-3
113-587-2070	Operate SINCGARS Single-Channel (SC)	AIT	AN	1-3
113-587-2071	Operate SINCGARS Frequency Hopping (FH) (Net Members)	AIT	AN	1-3
113-587-2072	Operate SINCGARS Frequency Hopping (FH) Net Control Station (NCS)	AIT	AN	1-3
113-587-2077	Operate SINCGARS Remote Control Unit (SRCU)	AIT	AN	1-3
113-587-0070	Troubleshoot Secure Single-Channel Ground and Airborne Radio Systems (SINCGARS) ICOM With or Without the AN/VIC-1 or AN/VIC-3	AIT	AN	1-3
113-589-0045	Troubleshoot Single-Channel Demand Assigned Multiple Access (DAMA) Tactical Satellite Terminal AN/PSC-5	AIT	AN	1-3
113-589-2028	Operate Secure Single-Channel Tactical Satellite Radio Set in SATCOM Mode	AIT	AN	1-3
113-589-2029	Operate Secure Single-Channel Tactical Satellite Radio Set in Demand Assigned Multiple Access (DAMA) Mode	AIT	AN	1-3
113-589-3045	Perform Scheduled Unit Level Maintenance (ULM) on Single-Channel Demand Assigned Multiple Access (DAMA) Tactical Satellite Terminal AN/PSC-5	AIT	AN	1-3
113-609-2053	Operate Automated Net Control Device (ANCD) AN/CYZ-10	AIT	AN	1-3

STP 11-25C13-SM-TG

CRITICAL TASKS

Task Number	Title	Training Location	Sust Tng Freq	Sust Tng SL
113-609-2052	Perform Net Control Station (NCS) Duties Using Automated Net Control Device (ANCD) AN/CYZ-10	AIT	AN	1-3
113-610-2044	Navigate Using the AN/PSN-11	AIT	AN	1-3
Subject Area 7. AUTOMATION OPERATIONS				
113-580-1032	Configure a Desktop IBM or Compatible Microcomputer for Operation	AIT	AN	1-3
113-580-1033	Install Network Hardware/Software in a Desktop IBM or Compatible Microcomputer	AIT	AN	1-3
113-580-1035	Install a Tactical Local Area Network (LAN)	AIT	AN	1-3
Subject Area 8. REMOTE COMMUNICATIONS				
113-622-1006	Install Radio Set Control Group AN/GRA-39(*)	UNIT	AN	1-3
113-622-2004	Operate Radio Set Control Group AN/GRA-39(*)	UNIT	AN	1-3
113-622-3001	Perform Unit Level Maintenance on Radio Set Control Group AN/GRA-39(*)	UNIT	AN	1-3
Subject Area 14. ENHANCED POSITION LOCATION REPORTING SYSTEM				
113-601-2056	Operate Generator Set MEP-803A	ASI	AN	1-3
113-630-1002	Prepare the EPLRS Net Control Station for Movement	ASI	AN	1-3
113-630-1005	Install the EPLRS Net Control Station	ASI	AN	1-3
113-630-1006	Deploy the EPLRS Gateway Radio Set	ASI	AN	1-3
113-630-1007	Deploy the Grid Reference Radio Set	ASI	AN	1-3
113-630-2001	Operate the EPLRS Net Control Station	ASI	AN	1-3
113-630-2027	Perform ECCM Procedures	ASI	AN	1-3
113-630-2031	Perform Playback Mode Operational Procedures	ASI	AN	1-3
113-630-2044	Operate the EPLRS Radio Set	ASI	AN	1-3
113-630-2045	Perform Multiple Net Control Station (NCS) Community Operational Procedures	ASI	AN	1-3
113-630-3011	Perform Preventive Maintenance Checks and Services (PMCS) on the EPLRS Net Control Station	ASI	AN	1-3
113-630-3014	Troubleshoot the EPLRS Net Control Station	ASI	AN	1-3
Subject Area 15. ABCS CORE				
171-580-1055	INSTALL FORCE XXI BATTLE COMMAND BRIGADE AND BELOW (FBCB2)	UNIT	AN	1-3
171-147-0001	PREPARE/SEND COMBAT MESSAGES USING FBCB2 VERSION 3.4	AIT	QT	1-3
171-147-0002	PERFORM STARTUP PROCEDURES FOR FORCE XXI BATTLE COMMAND BRIGADE AND BELOW (FBCB2) VERSION 3.4	AIT	AN	1-3
171-147-0005	APPLY MESSAGE ADDRESSING FEATURES IN FBCB2 VERSION 3.4	AIT	QT	1-4
171-147-0006	PERFORM MESSAGE MANAGEMENT USING FBCB2 VERSION 3.4	AIT	QT	1-4
171-147-0007	PREPARE/SEND OVERLAYS USING FBCB2 VERSION 3.4	AIT	QT	1-4
171-147-0008	PREPARE/SEND REPORTS USING FBCB2 VERSION 3.4	AIT	QT	1-4

16 February 2005

CRITICAL TASKS

Task Number	Title	Training Location	Sust Tng Freq	Sust Tng SL
171-147-0009	PREPARE/SEND FIRE/ALERT MESSAGES USING FBCB2 VERSION 3.4	AIT	QT	1-4
171-147-0010	PREPARE/SEND ORDER/REQUEST MESSAGES USING FBCB2 VERSION 3.4	AIT	QT	1-4
171-147-0011	PERFORM BEFORE-OPERATIONS PREVENTIVE MAINTENANCE CHECKS AND SERVICES ON FBCB2 VERSION 3.4	AIT	QT	1-4
171-147-0012	PERFORM SHUT-DOWN PROCEDURES FOR FBCB2 VERSION 3.4	AIT	QT	1-4
171-147-0013	PERFORM DURING-OPERATIONS PREVENTIVE MAINTENANCE CHECKS AND SERVICES ON FBCB2 VERSION 3.4	AIT	QT	1-4
171-147-0014	PERFORM AFTER-OPERATIONS PREVENTIVE MAINTENANCE CHECKS AND SERVICES ON FBCB2 VERSION 3.4	AIT	QT	1-4
171-147-0015	PREPARE/SEND A LOGISTICAL STATUS REPORT USING FBCB2 VERSION 3.4	AIT	QT	1-4
171-147-0017	EMPLOY MAP FUNCTIONS USING FBCB2 VERSION 3.4	AIT	QT	1-4
171-147-0019	EMPLOY FIPR FUNCTIONS USING FBCB2 VERSION 3.4	AIT	QT	1-4
171-147-0020	EMPLOY STATUS FUNCTIONS USING FBCB2 VERSION 3.4	AIT	QT	1-4
171-147-0021	EMPLOY ADMIN FUNCTIONS USING FBCB2 VERSION 3.4	AIT	QT	1-4
171-147-0022	EMPLOY APPS FUNCTIONS USING FBCB2 VERSION 3.4	AIT	QT	1-4
171-147-0023	EMPLOY NAV FUNCTIONS USING FBCB2 VERSION 3.4	AIT	QT	1-4
171-147-0024	EMPLOY QUICK SEND FUNCTIONS USING FBCB2 VERSION 3.4	AIT	QT	1-4
171-147-0025	EMPLOY FILTERS FUNCTIONS USING FBCB2 VERSION 3.4	AIT	QT	1-4
Skill Level 2				
Subject Area 9. **COMBAT COMMUNICATIONS PLANNING**				
113-587-7133	Direct Implementation of an FM Voice Data Communications Network	BNCOC	AN	2-3
113-589-7120	Direct Implementation of a Single-Channel Tactical Satellite Communications Network	BNCOC	AN	2-3
113-611-1001	Select Team Radio Site	BNCOC	AN	2-3
113-611-6112	Prepare Input to Signal Annex (Paragraph 5) of the OPORD	BNCOC	AN	2-3
113-620-7089	Direct Implementation of a High Frequency (HF) Communications Network	BNCOC	AN	2-3
Subject Area 12. **CONSTRUCT ANTENNAS**				
113-596-1098	Construct Field Expedient Antennas	UNIT	SA	2-3
Subject Area 13. **AUTOMATION OPERATIONS**				
113-580-0056	Troubleshoot Local Area Network (LAN)	BNCOC	AN	2-3

STP 11-25C13-SM-TG

CRITICAL TASKS

Task Number	Title	Training Location	Sust Tng Freq	Sust Tng SL
Skill Level 3				
Subject Area 9. COMBAT COMMUNICATIONS PLANNING				
113-611-6002	Plan FM Voice and Data Communications Net	BNCOC	AN	3
113-611-6004	Plan a Single-Channel Tactical Satellite Communications Network	BNCOC	AN	3
113-611-6005	Plan an HF Communications Network	BNCOC	AN	3
Subject Area 10. RADIO NET OPERATION				
113-571-7002	Inspect Station/Net Operation and Duties	UNIT	AN	2-3
113-573-1004	Conduct Communications Security Inspections	UNIT	AN	2-3
113-596-7081	Inspect Installation of Antennas	UNIT	AN	2-3
113-620-7088	Inspect Installed Operational Radio Sets	UNIT	AN	2-3
Subject Area 11. GENERATORS AND EQUIPMENT PREVENTATIVE MAINTENANCE CHECKS AND SERVICES				
113-601-7055	Inspect Installed Operational Generator Sets	UNIT	AN	2-3
113-617-7119	Inspect the Preventive Maintenance Checks and Services (PMCS)	UNIT	AN	2-3

16 February 2005

This page intentionally left blank.

STP 11-25C13-SM-TG

CHAPTER 3

MOS/Skill Level Tasks

Skill Level 1

Subject Area 1: RADIO PROCEDURES/MESSAGE HANDLING

Communicate in a Radio Net
113-571-1003

Conditions: Given a radio set, ACP 124(D), ACP 125 US Suppl-1, ACP 125(E), ACP 126(C), FM 24-1, FM 24-19, FM 24-35, and unit SOI.

NOTE: Supervision and assistance are available.

Standards: Established, entered, and left the radio net IAW the appropriate references.

Performance Steps
(Refer to ACP 124(D), ACP 125(E), ACP 125 US Suppl-1, ACP 126(C), and FM 24-35 for performance steps 1 through 3, except as noted.)

1. Establish a radio net
   a. Extract appropriate call signs, suffixes, and frequency from the unit SOI.
   b. Prepare and operate the appropriate radio set.
   c. Identify the net structure, determine the answering sequence, and make the appropriate response to the individual stations.

2. Enter a radio net. (Refer to ACP 124(D), ACP 125(E), ACP 126(C), and FM 24-1.)
   a. Use abbreviated call signs except when directed by NCS to use full call signs when confusion may result or when entering a net you do not normally operate in.
   b. Authenticate when challenged by NCS.
   c. If you fail to answer a multiple or collective call sign in sequence, wait until all other stations have answered, then answer. (Refer to FM 24-19.)
   d. If you are unable to communicate with NCS due to faulty equipment, wrong codes, unsuitable location, and so on, you must render a report to NCS as soon as possible by means other than a radio.

3. Stations leave and close the net.
   a. Request permission from NCS to leave the net.
   b. Inform NCS of the reason for leaving the net.
   c. Authenticate upon direction of NCS prior to leaving the net.

Performance Measures	GO	NO GO
(Refer to ACP 124(D), ACP 125 (E), ACP 125 US Suppl-1, ACP 126(C), and FM 24-35 for PMs 1 through 3, except as noted.)		
1. Established a radio net.	——	——
2. Entered a radio net. (Refer to ACP 124(D), ACP 125(E), ACP 126(C), and FM 24-1.)	——	——
3. Stations left and closed the net.	——	——

STP 11-25C13-SM-TG

Evaluation Guidance: Score the Soldier a GO if all PMs are passed. Score the Soldier a NO-GO if any PM is failed. If the Soldier fails any PM, show what was done wrong and how to do it correctly. Have the Soldier perform the PMs until they are done correctly.

References

Required	**Related**
ACP 124(D)	FM 24-18
ACP 125 US SUPPL-1	FM 24-35-1
ACP 125(E)	
ACP 126(C)	
FM 24-1	
FM 24-19	
FM 24-35	
UNIT SOI	

STP 11-25C13-SM-TG

Send a Message in 16-Line Format
113-572-1002

Conditions: Given ACP 124(D), ACP 125(E), ACP 126(C), and DA PAM 25-7.

NOTE: Supervision and assistance are available.

Standards: Prepared and transmitted a given message in the proper message format.

Performance Steps

1. Prepare given message in correct format for transmission. (Refer to ACP 124(D), ACP 125(E), ACP 126(C), and DA PAM 25-7.)
    a. Radiotelephone, USMTF.
    b. Radio teletypewriter, USMTF (if required).
    c. Radiotelegraph.

2. Ensure correct message format is used for mode of transmission. (Refer to ACP 124(D), ACP 125(E), ACP 126(C), and DA PAM 25-7.)
    a. Radiotelephone.
    b. Radioteletypewriter.
    c. Radiotelegraph.

Evaluation Preparation: Setup: For this evaluation, prepare message in correct format for transmission.

Brief Soldier: Tell the Soldier that performance measures (PMs) 1a and 1b (1c for A-4 qualified personnel) and 2a and 2b (2c for A-4 qualified personnel) must be passed.

Performance Measures	GO	NO GO
1. Prepared given message in correct format for transmission. (Refer to ACP 124(D), ACP 125(E), ACP 126(C), and DA PAM 25-7.)	——	——
2. Ensured correct message format was used for mode of transmission. (Refer to ACP 124(D), ACP 125(E), ACP 126(C), and DA PAM 25-7.)	——	——

Evaluation Guidance: Score the Soldier a GO if all PMs are passed. Score the Soldier a NO-GO if any PM is failed. If the Soldier fails any PM, show what was done wrong and how to do it correctly. Have the Soldier perform the PMs until they are done correctly.

References

Required	Related
ACP 124(D)	
ACP 125(E)	
ACP 126(C)	
DA PAM 25-7	

STP 11-25C13-SM-TG

Recognize Electronic Attack (EA) and Implement Electronic Protection (EP)
113-573-6001

Conditions: Given a radio set, applicable operator's technical manual, FM 24-1, FM 24-33, and unit SOI extract.

Standards: Determined that electronic warfare (EW) is directed at your station and electronic counter-countermeasures (ECCM) are employed for continued operation.

Performance Steps

1. Introduction. A close relationship exists between ECCM and COMSEC. Both defensive arts are based on the same principle. An enemy who does not have access to our essential elements of friendly information (EEFI) is a much less effective foe. The major goal of COMSEC is to ensure that friendly use of the electromagnetic spectrum for communications is by the enemy. The major goal of practicing sound ECCM techniques is to ensure the continued use of the electromagnetic spectrum. ECCM techniques are designed to ensure commanders some degree of confidence in the continued use of these techniques. Our objective must be to ensure that all communications equipment can be employed effectively by tactical commanders in spite of the enemy's concerted efforts to degrade such communications to the enemy's tactical advantage. The modification and the development of equipment to make our communications less susceptible to enemy exploitation are expensive processes. Equipment is being developed and fielded which will provide an answer to some of ECCM problems. Commanders, staff, planners, and operators remain responsible for security and continued operation of all communications equipment.

    a. Operators of communications equipment must be taught what jamming and deception can do to communications. They must be made aware that incorrect operating procedures can jeopardize the unit's mission and ultimately increase unit casualties. Preventive and remedial ECCM techniques must be employed instinctively. Maintenance personnel must be made aware that unauthorized or improperly applied modifications may cause equipment to develop peculiar characteristics, which can be readily identified by the enemy.

    b. ECCM should be preventive in nature. ECCM should be planned and applied to force the enemy to commit more jamming, interception and deception resources to a target than it is worth, or is available. ECCM techniques must also be applied to force the enemy to doubt the effectiveness of the enemy's jamming and deception efforts.

    c. Before we can begin to prevent electronic countermeasures (ECM), we must first be certain of what we are trying to prevent.

        (1) Jamming is the deliberate radiation, re-radiation, or reflection of electromagnetic energy with the object of impairing the use of electronic devices, equipment, or systems. The enemy conducts jamming operations against us to prevent us from effectively employing our radios, radars, navigational aids (NAVAIDS), satellites, and electro-optics. Obvious jamming is normally very simple to detect. The more commonly used jamming signals of this type are described below. Do not try to memorize them; just be aware that these and others exist. When experiencing a jamming incident, it is much more important to recognize it and take action to overcome it than to identify it formally.

            (a) Random noise. It is random in amplitude and frequency. It is similar to normal background noise and can be used to degrade all types of signals.

            (b) Stepped tones. These are tones transmitted in increasing and decreasing pitch. They resemble the sound of bagpipes.

            (c) Spark. The spark is easily produced and is one of the most effective forms of jamming. Bursts are of short duration and high intensity. Sparks are repeated at a rapid rate and are effective in disrupting all types of communications.

            (d) Gulls. The gull signal is generated by a quick rise and a slow fall of a variable radio frequency and is similar to the cry of a sea gull.

Performance Steps
- (e) Random pulse. In this type of interference, pulses of varying amplitude, duration, and rate are generated and transmitted. Random pulses are used to disrupt teletypewriter, radar, and all types of data transmission systems.
- (f) Wobbler. The wobbler is a single frequency, which is modulated by a low and slowly varying tone. The result is a howling sound, which causes a nuisance on voice radio communications.
- (g) Recorded sounds. Any audible sound, especially of a variable nature, can be used to distract radio operators and disrupt communications. Examples of sounds include: music, screams, applause whistles, machinery noise, and laughter.
- (h) Preamble jamming. This type of jamming occurs when the synchronization tone of speech security equipment is broadcast over the operating frequency of secure radio sets. Preamble jamming results in radios being locked in the receive mode. It is especially effective when employed against radio nets using speech security devices.
- (i) Subtle jamming. This type of jamming is not obvious at all. With subtle jamming, no sound is heard from our receivers. They cannot receive incoming friendly signals, but everybody appears normal to the radio operator.

(2) Meaconing. This is a system of receiving radio beacon signals from NAVAIDS and rebroadcasting them on the same frequency to confuse navigation. The enemy conducts meaconing operations against us to prevent our ships and aircraft from arriving at their intended targets or destinations.

(3) Intrusion. Intentional insertion of electromagnetic energy into transmission paths with the objective of deceiving equipment operators or causing confusion. The enemy conducts intrusion operations against us by inserting false information into our receiver paths. This false information may consist of voice instructions, ghost targets, coordinates for fire missions, or even rebroadcasting of prerecorded data transmissions.

(4) Interference. Interference is any electrical disturbance, which causes undesirable responses in electronic equipment. As a MIJI term, interference refers to the unintentional disruption of the use of radios, radars, NAVAIDS, satellites, and electro-optics. This interference may be of friendly, enemy, or atmospheric origin. For example, a civilian radio broadcast interrupting military communications is interference.

2. Communications Protective Measures.
   a. Considerations. Properly applied ECCM techniques will deny valuable intelligence sources to the enemy and eliminate much of the threat that he poses to our combat operations. The following discussion describes practical ways to protect communications systems.
   b. The siting of the transmitting antenna is critical in the ECCM process. Before making a decision about a proposed site for either a single-channel or multichannel antenna, there are two basic questions to answer:
      (1) Are communications possible from the proposed site?
      (2) Are there enough natural obstacles between the site and the enemy to mask transmission?
   c. The final decision on site selection will often be a tradeoff between the answers to these two questions. The communications mission must have first priority in determining the actual antenna sites. There are additional actions that must be taken to limit the enemy's chances of interception and location successes. Transmitters and antennas should be located away from the headquarters. The two locations should be separated by more than 1 kilometer (0.62 mile). Erroneous radio frequency direction (RFD) data used in conjunction with observation data may favor the targeting of a decoy site instead of the actual transmitter site. This ploy depends upon good camouflage at the actual site. Transmitters grouped in one area indicate the relative value of the headquarters. Directional antennas reduce radiation exposure to enemy receivers and enhance the intended signal. (For instruction on directional antennas, refer to TC 24-21.)
   d. Use the lowest possible transmitter power output. Power means less radiated power reaches the enemy and thus increases his difficulty in applying ECM.

STP 11-25C13-SM-TG

Performance Steps
  e. Use only approved code systems. Never use unauthorized (homemade) codes. Use of non-NSA generated codes can provide a false COMSEC sense of security that can be exploited by enemy radio intercept operators. Only when absolutely necessary should traffic be passed in the clear.
  f. Rather than assuming equipment is defective, assume that it is operational. Operators must not contact other stations for equipment checks simply because no message has been transmitted in a set time frame.

Evaluation Preparation: Setup: A radio set operating in a radio net with interference applied to the system.

Brief Soldier. Tell the Soldier to apply proper tactics to the jamming system.

Performance Measures   **GO**   **NO GO**

1. Determined if ECM was being employed.
   a. Checked for accidental or unintentional interference. (Refer to FM 24-33.)
   b. Checked for intentional interference. (Refer to FM 24-33.)

2. Initiated operator's procedures. (Refer to FM 24-1 and FM 24-33.)
   a. Checked the equipment ground to ensure that the interference was not caused by a buildup of static electricity.
   b. Disconnected the antenna.
   c. Identified the type of sound.
   d. Moved the receiver or reoriented the antenna, if possible, and listened or looked for variations in the strength of the disturbance.
   e. Tuned the receiver above or below the normal frequency. If such detuning caused the intensity of the interfering signal to drop sharply, it can be assumed that the interference was the result of spot jamming.

3. Identified jamming signals. (Refer to FM 24-33.)

4. Employed antijamming measures. (Refer to FM 24-1.)

NOTE: Antijamming measures are designed to allow radio operators to work effectively through intentional interference. Regardless of the nature of the interfering signal, radio operators WILL NOT reveal in the clear the possibility or success of enemy jamming.

Evaluation Guidance: Score the Soldier a GO if all PMs are passed. Score the Soldier a NO-GO if any PM is failed. If the Soldier fails any PM, show what was done wrong and how to do it correctly. Have the Soldier perform the PMs until they are done correctly.

References
  Required   **Related**
  FM 24-1   FM 24-18
  FM 24-33   TC 24-21
  UNIT SOI

STP 11-25C13-SM-TG

Submit Interference Message Report
113-573-7017

Conditions: You will be required to submit an interference message report. This task is performed in any condition or situation. Given FM 24-33, DA PAM 25-7, and unit SOI.

NOTE: Supervision and assistance are available.

Situation I. While performing the duties of an operator in a radio net/system, you encounter an interfering signal of unknown origin, possibly an MIJI incident.

Situation II. One of your duties as a radio operator is to prepare and submit MIJI incident reports.

Standards: Prepared and submitted the MIJIFEEDER voice template message report through the NCS by the best available means and assembled the supplemental information for the follow-up MIJI report.

Performance Steps

1. Introduction. Meaconing, intrusion, and jamming are deliberate actions intended to deny any enemy the effective use of the electromagnetic spectrum. Interference is unintentional disruption of the effective use of the electromagnetic spectrum by friendly, enemy, or atmospheric sources. Collectively, meaconing, intrusion, jamming, and interference incidents are referred to as MIJI incidents.

2. MIJI reports document all disruptions of the following:
    a. Radios.
    b. Radars.
    c. Navigational aids (NAVAIDS).
    d. Satellites.
    e. Electro-optics.

3. Disruptions that are caused by equipment malfunctions or destruction are exceptions. The MIJI report serves two purposes. First, it provides information to the tactical commander allowing timely decisions to be made to overcome the MIJI problem. Second, it provides a historical record of MIJI incidents from which appropriate ECCM techniques and measures can be developed. This helps us to counter future attempts by the enemy to deny us the effective use of the electromagnetic spectrum.

4. This task gives instructions for completing MIJI reports for communications and noncommunications emitters. To fulfill the two purposes stated above, there are two kinds of MIJI reports. The MIJIFEEDER voice template message is a brief report of a MIJI incident. It serves as a decision making tool for the command. The MIJIFEEDER record message is a complete report of a MIJI incident. This provides a historical record from which appropriate ECCM techniques and measures can be developed. To find the definition of meaconing, intrusion, jamming, and interference, see Training Information Outline of Task 113-573-6001, Recognize Electronic Attack (EA), and Implement Electronic Protection (EP).

5. Purpose and Use of the MIJIFEEDER Voice Template. It is only the information needed to adequately inform the tactical commander of the incident in a timely manner. It is used to make evaluation of enemy actions or intentions easier and to provide data to implement appropriate counter-countermeasures. The equipment operator experiencing the MIJI incident forwards it through the chain of command to the unit operations center. The report should be forwarded using the most expeditious service communications means available.

STP 11-25C13-SM-TG

Performance Steps

6. Report Format and Use of the MIJIFEEDER Voice Template. The voice template was developed for use under the JINTACCS program. It is designed to ensure interoperability on the battlefield during combined, joint, and inter-Army operations. The standardized, simple format permits the expeditious notification of appropriate action elements in time-critical situations. Only the completed and underlined areas (as appropriate) of the format are transmitted. The MIJIFEEDER voice templates are self-explanatory and contain 10 items of information. When the message is transmitted over nonsecure means, each line number is stated and the completed information is encrypted. When a secure means is used, the title of each line is transmitted along with the completed information.

7. Purpose and use of the MIJIFEEDER Voice Template. The MIJIFEEDER record is a complete report of a MIJI incident. It provides a basis for developing appropriate counteraction measures to be implemented at proper command levels. The Joint Electronic Warfare Center (JEWC) is the action agency for this report. All MIJIFEEDER record message reports initially evaluated as nonexercise should be forwarded as soon as possible to the JEWC. The JEWC uses these reports to develop trends and to evaluate foreign ECM operations. They are also used by the JEWC to recommend operational methods and equipment changes that will reduce MIJI vulnerability of radios, radars, NAVAIDS, satellites, and electro-optics.

8. Reporting Procedures of the MIJIFEEDER Record Message. The Signal officer of the affected unit forwards the message to the JEWC, OPM, San Antonio, TX, through operations channels to the corps operation center. All MIJIFEEDER reports are forwarded via secure means within 24 hours of the MIJI incident. Security classification of MIJI incidents or evaluation reports is determined principally by intent and location of the implied or stated source of the problem. Stations in combat areas or having a sensitive military mission ordinarily classify all MIJI reports.

9. Report Format and Contents. In order to complete this report, you must use FM 24-33 and DA PAM 25-7.

Evaluation Preparation: Setup: You are provided with FM 24-33 and DA PAM 25-7.

Brief Soldier: You must process a MIJIFEEDER Voice Template/MIJIFEEDER Record Message Report in proper format.

Performance Measures	**GO**	**NO GO**
1. Entered the unit designation.	——	——
2. Entered the type of interference encountered.	——	——
3. Entered the unit location in either of two ways: longitude in minutes and seconds, or in complete grid coordinates down to 10- or 100-meter increments.	——	——
4. Entered two digits each for day, hour, and minutes and one letter for the time zone for the start of the MIJI incident.	——	——
5. Entered two digits each for day, hour, minutes, and one letter for the time zone for the end of the MIJI incident.	——	——
6. Entered the nomenclature for the equipment affected.	——	——
7. Entered the channel, frequency or frequency range affected, and the unit of measure.	——	——
8. Entered, in his or her own words, a brief description or other information regarding the MIJI incident.	——	——
9. When required, entered the hours, minutes, and time zone.	——	——

Performance Measures <u>GO</u>   <u>NO GO</u>

10. Entered the message authentication IAW the joint task form (JTF) requirements.   ___   ___

Evaluation Guidance: Score the Soldier a GO if all PMs are passed. Score the Soldier a NO-GO if any PM is failed. If the Soldier fails any PM, show what was done wrong and how to do it correctly. Have the Soldier perform the PMs until they are done correctly.

References
    Required   **Related**
    DA PAM 25-7
    FM 24-33
    UNIT SOI

STP 11-25C13-SM-TG

Operate VIASAT Software
113-572-1003

Conditions: In a field or classroom environment, given an Airborne Data Controller (ADC) VDC-300, VDC-300 operator's manual (OM), data terminal loaded with ViaSat's Data Terminal Software for Windows (DTS/Win) or E-mail Software (EMAIL), ViaSat E-mail User's Guide (UG), crypto device, power supply, and radio set or telephone.

Standards: Transferred data using ViaSat software via the VDC-300 over secure communications channels.

Performance Steps

1. Apply power to the ADC VDC-300.
    a. Ensure the external power supply is attached to the rear of the unit.
    b. Turn the ON/OFF switch clockwise to the ON position. The switch has four positions.

NOTE: When the switch is turned from the OFF position to the ON position, the LEDs and all LCD segments will be illuminated for approximately three seconds. Following this power-on cycle, the LEDs will extinguish and the VDC-300 will be ready for normal operation. If the LEDs remain on, the VDC-300 has detected a failure. (Return to VIASAT for repair).

    (1) (BIT): The unit continuously runs a built-in test.

NOTE: When the switch is turned from the OFF position to the BIT position, the VDC-300 will continuously perform built-in test. If a failure occurs, one or more LEDs will light and remain on.

    (2) (OFF): The unit is powered OFF.
    (3) (BLK): Blackout Mode. The unit is powered ON, ready for operation using DTS. The LEDs will not light for night vision goggle compatibility.
    (4) (ON): The unit is powered ON, ready for operation using DTS. The LEDs will be functional.

2. Configure the VDC-300 for operation with the crypto device.
    a. Ensure crypto device is connected to VDC-300.
    b. Turn the Crypto Select Switch to the position that corresponds to the crypto device in use. (Refer to the operator's manual.)

3. Start ViaSat E-mail.

NOTE: Menu selection options are abbreviated as Menu / Selection. For example, "choose File / Print" means click on the File menu and select Print from the drop-down menu box. References to words or text in a window (such as window or dialog box titles, labels, buttons, frame text, list box name, or list choices) are in Initial Caps.

    a. Click Start button on Windows Taskbar.
    b. Choose Programs / ViaSat / ViaSat E-mail; this will bring up the Main Application Window.

4. Send E-mail.
    a. Compose E-mail.
        (1) Click the Send New E-mail button on the toolbar or choose Message / Send New E-mail. This brings up the Composing New E-mail screen.
        (2) Address the message, type in a known address in the To field, or click on the Address button to access the E-mail Addressed To window.

NOTE: To address your message using the E-mail Addressed To window, select destinations in the left column and click the Add button to place the destination addresses in the Selected Addresses column. When you are finished selecting addresses, click Done. This returns you to the Composing New E-mail screen.

Performance Steps
- (3) Enter a subject line in the Subject edit box, and then enter your message in the message composition box.
- (4) Set the send priority option. (Routine, Priority, or Immediate.)
- (5) Set the classification level. (Unclassified, Confidential, Secret, or Top Secret.)
- (6) Set SEND option. (Compress, Return Receipt, or Reliable.)

b. Send your message. (Click the Send button.)

5. Receive E-mail.
  a. Read incoming messages.
   - (1) Click on the E-mail Folders button on the toolbar or choose Message / E-mail. This brings up the E-mail Folders window.
   - (2) Click on the Inbox in the folders list. Received messages will appear in the Inbox.

NOTE: New (unread) E-mail messages will prefix with the NEW icon.

- (3) To view the message in a separate window, double-click the message in the message list.

Performance Measures	**GO**	**NO GO**
1. Applied power to the ADC VDC-300.	——	——
2. Configured the VDC-300 for operation with the crypto device.	——	——
3. Started ViaSat E-mail.	——	——
4. Sent E-mail.	——	——
5. Received E-mail.	——	——

Evaluation Guidance: Score the Soldier a GO if all PMs are passed. Score the Soldier a NO-GO if any PM is failed. If the Soldier fails any PM, show what was done wrong and how to do it correctly. Have the Soldier perform the PMs until they are done correctly.

References
  Required   **Related**
  VDC-300 OM
  VIASAT UG

STP 11-25C13-SM-TG

Subject Area 2: ANTENNAS

Construct Vertical Half-Rhombic Antenna
113-596-1052

Conditions: Given 200 feet of W-1 antenna wire, electrical tape, 400 to 600 ohms terminating resistor, insulators or material to construct field expedient insulators, radio set, knife, suspension line, measuring device, compass, SOI with frequency and call signs, suitable training site, FM 24-18, and FM 11-487-1.

Standards: Constructed a vertical half-rhombic antenna and it was operational.

Performance Steps

1. Construct antenna.

NOTE: Counterpoise should be the length from leg closest to distant station to leg furthest from distant station.

    a. Measure 100 feet of antenna wire total length. (Add 12 inches to connect insulators.)
    b. Bend wire in half to find apex point for connection of insulator.
    c. Connect insulator to apex point.
    d. Connect halyard to insulator at apex point.
    e. Tape wire to insulator to prevent movement to apex point.

2. Erect antenna.
    a. Determine azimuth to distant station.
    b. Install antenna so apex is at least 30 feet high.
    c. Separate legs and equal distance on azimuth to distant station.
    d. Install counterpoise.
    e. Connect terminating resistor.

NOTE: One end of resistor is connected to end of antenna; the other end is connected to counterpoise and closet to distant station.

3. Connect feeder line.
    a. Connect one wire to antenna and positive side to radio.
    b. Connect other wire from end of counterpoise to ground on radio.

4. Call distant station.

Performance Measures	**GO**	**NO GO**
1. Constructed antenna.	——	——
2. Erected antenna.	——	——
3. Connected feeder line.	——	——
4. Called distant station.	——	——

Evaluation Guidance: Score the Soldier a GO if all PMs are passed. Score the Soldier a NO-GO if any PM is failed. If the Soldier fails any PM, show what was done wrong and how to do it correctly. Have the Soldier perform the PMs until they are done correctly.

References
   Required   **Related**
   FM 11-487-1
   FM 24-18
   UNIT SOI

STP 11-25C13-SM-TG

Construct a Long-Wire Antenna
113-596-1056

Conditions: Given a suitable area, paper, pencil, antenna wire, three insulators, a 50-foot guy rope, anchor stakes, hammer, knife, pliers, measuring tape, compass, frequency, a suitable radio, azimuth to orient the antenna, FM 24-18, and FM 11-487-1.

NOTE: The AN/PRC-74 radio set is not a suitable radio due to the impedance mismatch.

Standards: Constructed and erected the antenna so it is within ±3 inches of the required length.

Performance Steps

1. Compute antenna length by using the following formula: L= 492 (N - .05)) ÷ Frequency (MHz); L = length in feet; N = number of half wavelengths; (1 wavelength = 2 half wavelengths).

EXAMPLE: Build a 4-wavelength antenna for a frequency of 12 MHz. L = 492 (N - .05) ÷ Frequency; L = 492 (8 - .05) ÷ 12; L = 492 x 7.95 ÷ 12; L = 326 feet.

2. Assemble antenna.
    a. Measure antenna wire using computations from performance step 1.
    b. Attach insulators.

3. Install antenna.
    a. Using the compass, orient antenna to the direction of maximum desired radiation. (Long wires can be directional antennas.)
    b. Erect antenna.

Performance Measures	**GO**	**NO GO**
1. Computed antenna length by using the formula.	——	——
2. Assembled antenna.	——	——
3. Installed antenna.	——	——

Evaluation Guidance: Score the Soldier a GO if all PMs are passed. Score the Soldier a NO-GO if any PM is failed. If the Soldier fails any PM, show what was done wrong and how to do it correctly. Have the Soldier perform the PMs until they are done correctly.

References

Required	Related
FM 11-487-1	
FM 24-18	

Install Antenna Group OE-254/GRC (Team Method)
113-596-1068

Conditions: In a tactical or nontactical situation, given antenna group OE-254/GRC, two persons for erecting the antenna, frequency modulation (FM) radio set (installed), DA PAM 738-750, and TM 11-5985-357-13.

Standards: Installed and connected the OE-254/GRC to the FM radio set and performed PMCS IAW TM 11-5985-357-13 and DA PAM 738-750 within 25 minutes.

Performance Steps
(Refer to TM 11-5985-357-13 for all performance steps.)

1. Perform PMCS.
2. Plan antenna installation site.
3. Position baseplate and guy stakes.
4. Assemble antenna equipment.
5. Erect antenna using two persons.
6. Connect the CG-1889B/U connector to the radio.

Performance Measures (Refer to TM 11-5985-357-13 for all PMs.)	<u>GO</u>	<u>NO GO</u>
1. Performed PMCS.	——	——
2. Planned antenna installation site.	——	——
3. Positioned baseplate and guy stakes.	——	——
4. Assembled antenna equipment.	——	——
5. Erected antenna using two persons.	——	——
6. Connected the CG-1889B/U connector to the radio.	——	——

Evaluation Guidance: Score the Soldier a GO if all PMs are passed. Score the Soldier a NO-GO if any PM is failed. If the Soldier fails any PM, show what was done wrong and how to do it correctly. Have the Soldier perform the PMs until they are done correctly.

References

Required	**Related**
DA PAM 738-750	FM 24-18
TM 11-5985-357-13	GTA 11-3-20

STP 11-25C13-SM-TG

Construct a Doublet Antenna
113-596-1070

Conditions: Given a requirement and a radio set, mast AB/155(*)/U (three each), antenna group AN/GRA-50 or sufficient W-1 antenna wire for the construction of the doublet antenna to the assigned frequency, compass M-2 or equivalent, TM 11-5815-334-10, TM 11-5820-256-10, TM 11-5820-467-15, and wire cutter kit TE-33 or equivalent.

NOTE: Supervision and assistance are available.

Standards: Properly cut doublet antenna frequency and erected broadside to the most distant station.

Performance Steps

NOTE: Performance step 3 is a team task.

1. Construct antenna using W-1 antenna wire (refer to TM 11-5820-256-10, TM 11-5815-334-10, and TM 11-5820-467-15); or construct antenna using antenna group AN/GRA-50 (refer to TM 11-5820-467-15).
    a. Use formula 468 ÷ Frequency = Length.
    b. Frequency____MHz.

EXAMPLE: 468 ÷ 26.00 MHz = Length; 26 x 18 feet = 468; 18 feet half wave or 9 feet quarter wave center fed.

2. Prepare mast AB-155(*)/U for erection. (Refer to TM 11-5820-256-10 and TM 11-5815-334-10.)

3. Erect antenna. (Refer to TM 11-5820-256-10 and TM 11-5815-334-10.)
    a. Antenna must be broadside to the most distant station. Determine azimuth by using compass.
    b. Connect antenna lead-in to radio set. (Refer to TM 11-5820-256-10.)

Performance Measures	GO	NO GO

NOTE: PM 3 is a team task.

	GO	NO GO
1. Constructed antenna using W-1 antenna wire (refer to TM 11-5820-256-10, TM 11-5815-334-10, and TM 11-5820-467-15); or constructed antenna using antenna group AN/GRA-50 (refer to TM 11-5820-467-15 and Figures 3-4, 3-5, and 3-6).	——	——
2. Prepared mast AB-155(*)/U for erection. (Refer to TM 11-5820-256-10 and TM 11-5815-334-10.)	——	——
3. Erected antenna. (Refer to TM 11-5820-256-10 and TM 11-5815-334-10.)	——	——

Evaluation Guidance: Score the Soldier a GO if all PMs are passed. Score the Soldier a NO-GO if any PM is failed. If the Soldier fails any PM, show what was done wrong and how to do it correctly. Have the Soldier perform the PMs until they are done correctly.

References

Required	Related
TM 11-5815-334-10	TM 11-5820-520-10
TM 11-5820-256-10	
TM 11-5820-467-15	

STP 11-25C13-SM-TG

Construct a Near Vertical Incidence Skywave (NVIS) Antenna AS-2259/GRA
113-596-1071

Conditions: Given a complete antenna kit AS-2259/GRA, antenna adapter MX-10618, and TM 11-5820-919-12, TM 11-5820-923-12, or TM 11-5820-924-13.

Standards: Constructed and erected the antenna IAW technical manual references.

Performance Steps

DANGER: Antenna installation area must be free of power lines. Antenna contact with power lines during installation may cause serious injury or even DEATH to the operator.

1. Determine installation area. An 84 X 84 foot area is required for NVIS installation.

2. Place antenna adapter MX-10618 on the whip antenna mast base AB-652 and ground it with the ground strap located on the adapter.

3. Open the antenna pack and remove the topmast assembly.

4. Install the topmast assembly on the antenna adapter MX-10618.
   a. Uncoil the antenna elements one at a time.
   b. Verify that the antenna elements are stretched along the direction in which they leave the top housing and are NOT SHORTED to each other or to the mast.

5. Measure anchor positions, using metal sleeve cable markers as guides, and install stakes. Leave slack in elements lying on the ground.

6. Before connecting mast sections, wipe unpainted mating surfaces clean of mud or dirt to provide good electrical contact.

7. Assemble mast by raising the topmast assembly and insert only as many mast as are necessary to raise the top of the antenna about 16 feet from the ground. Place the last section onto the MX-10618.

8. Adjust tension on all elements until mast is vertical and straight. Elements need not be excessively taut (3 to 5 pounds of tension).

Performance Measures	GO	NO GO
DANGER: Antenna installation area must be free of power lines. Antenna contact with power lines during installation may cause serious injury or even DEATH to the operator.		
1. Determined installation area. An 84 X 84 foot area is required for NVIS installation.	——	——
2. Placed antenna adapter MX-10618 on the whip antenna mast base AB-652 and grounded it with the ground strap located on the adapter.	——	——
3. Opened the antenna pack and removed the topmast assembly.	——	——
4. Installed the topmast assembly on the antenna adapter MX-10618.	——	——
5. Measured anchor positions, using metal sleeve cable markers as guides, and installed stakes. Left slack in elements lying on the ground.	——	——
6. Before connecting mast sections, wiped unpainted mating surfaces clean of mud or dirt to provide good electrical contact.	——	——

STP 11-25C13-SM-TG

Performance Measures	GO	NO GO
7. Assembled mast by raising the topmast assembly and inserted only as many masts as are necessary to raise the top of the antenna about 16 feet from the ground. Placed the last section onto the MX-10618.	——	——
8. Adjusted tension on all elements until mast was vertical and straight. Elements need not be excessively taut (3 to 5 pounds of tension).	——	——

Evaluation Guidance: Score the Soldier a GO if all PMs are passed. Score the Soldier a NO-GO if any PM is failed. If the Soldier fails any PM, show what was done wrong and how to do it correctly. Have the Soldier perform the PMs until they are done correctly.

References
    Required   **Related**
    TM 11-5820-919-12   TM 11-5985-379-14&P
    TM 11-5820-923-12
    TM 11-5820-924-13

STP 11-25C13-SM-TG

Subject Area 3: DATA COMMUNICATIONS EQUIPMENT

Install Terminal Communications AN/UGC-74B(V)3
113-598-1023

Conditions: Given terminal communications AN/UGC-74B(V)3, associated radio sets, and TM 11-5815-602-10.

NOTE: Supervision and assistance are available.

Standards: Installed terminal communications AN/UGC-74B(V)3 in the mount without damage, made cable connections, aligned the AN/UGC-74B(V)3, and conducted a local test of the system IAW PMs 1 through 3 within a 30-minute time limit.

Performance Steps
(Refer to TM 11-5815-602-10 for all performance steps.)

 1. Install terminal AN/UGC-74(V)3 in the mount.

 2. Install the plug-in items to the terminal connectors.

NOTE: The terminal connectors are located at the rear of the case. Access to the connectors is gained through the rear panel door. Open the door by pulling the door handle down into the horizontal position and rotating it 1/4 turn to the right. Secure the door in the open (raised) position by unsnapping the retaining strap from the outer case cover and inserting the rear panel door handle into the retaining strap slot. Ensure that sufficient slack remains in the cables after they are connected to allow for the extension of the machine from its case.

 3. Conduct a local test of the terminal.

Performance Measures	**GO**	**NO GO**
(Refer to TM 11-5815-602-10 for all PMs.)		
1. Installed terminal AN/UGC-74B(V)3 in the mount.	——	——
2. Installed the plug-in items to the terminal connectors.	——	——
3. Conducted a local test of the terminal.	——	——

Evaluation Guidance: Score the Soldier a GO if all PMs are passed. Score the Soldier a NO-GO if any PM is failed. If the Soldier fails any PM, show what was done wrong and how to do it correctly. Have the Soldier perform the PMs until they are done correctly.

References
Required	Related
TM 11-5815-602-10	

STP 11-25C13-SM-TG

Operate Terminal Communications AN/UGC-74B(V)3
113-598-2010

Conditions: Given terminal communications AN/UGC-74B(V)3, associated radio equipment, and TM 11-5815-602-10.

NOTE: Supervision and assistance are available.

Standards: Prepared the terminal for the selected operating state IAW PM 1 and placed in operation and taken out of operation IAW PMs 1 through 5 within a 20-minute time limit.

Performance Steps
(Refer to TM 11-5815-602-10 for all performance steps.)

1. Perform preliminary starting procedures.

2. Perform initial adjustment.

3. Prepare the terminal for operation in the RECEIVE ONLY (RO) state.

NOTE: Based on the initial adjustments made, the terminal will print out the operation validations/state determination message for the RO state. After the state determination/validation state message is printed, the terminal is ready to receive data.

CAUTION: If a self-test is to be performed, it will be initiated after the state determination/validation state message has printed out and before message data is stored in memory.

4. Prepare the terminal for operation in the KEYBOARD SEND/RECEIVE (KSR) state.

NOTE: The KSR state expands the capability of the terminal from the RO state by making the keyboard available to the operator. Messages are composed in the conventional manner. However, the terminal provides the operator with the capability of sending messages one print-line at a time, as opposed to the usual character-by-character technique. This allows the operator to compose, edit, and review a full 80-character line of message before transmission. Based on the initial adjustments made, the terminal will print out the operation validation/state determination for the KSR state. After the printout, the machine is in the KSR state and ready for message reception or transmission.

CAUTION: If a self-test is conducted, it will be performed immediately after the operation validation/state determination message has printed.

5. Prepare the terminal for operation in the INTELLIGENT COMMUNICATIONS TERMINAL (ICT) state.

NOTE: The ICT state provides the operator with composing, editing, and formatting capabilities. These capabilities are gained by using the system command structure. Operation in the ICT state should be performed IAW the same procedures as the RO and KSR states, except that the system is powered down. The STATE switch is placed in the ICT state and then powered up. Based on the initial adjustments made, the terminal will print out the operation validation/state determination message for the ICT state. Before placing the POWER switch to OFF, verity that the operation is complete and all printing has ceased.

6. Type the 25 wpm typing test within a 5-minute period with no more than five errors.

Evaluation Preparation: Setup: Ensure the equipment is complete and operational. Prepare a 25 wpm typing test to last exactly 5 minutes for PM 6.

STP 11-25C13-SM-TG

Brief Soldier: Tell the Soldier that in addition to performing PMs 1 through 5 within a 20-minute time limit, PM 6 must be completed within 5 minutes. PMs will be performed in sequence and must be passed.

Performance Measures   <u>GO</u>   <u>NO GO</u>
(Refer to TM 11-5815602-10 for all PMs.)

1. Performed preliminary starting procedures.   ——   ——

2. Performed initial adjustment.   ——   ——

3. Prepared the terminal for operation in the RECEIVE ONLY (RO) state.   ——   ——

4. Prepared the terminal for operation in the KEYBOARD SEND/RECEIVE (KSR) state.   ——   ——

5. Prepared the terminal for operation in the INTELLIGENT COMMUNICATIONS TERMINAL (ICT) state.   ——   ——

6. Typed the 25 wpm typing test within a 5-minute period with no more than five errors.   ——   ——

Evaluation Guidance: Score the Soldier a GO if all PMs are passed. Score the Soldier a NO-GO if any PM is failed. If the Soldier fails any PM, show what was done wrong and how to do it correctly. Have the Soldier perform the PMs until they are done correctly.

References
    Required   **Related**
    TM 11-5815-602-10

STP 11-25C13-SM-TG

Perform Unit Level Maintenance on Communications Terminal AN/UGC-74B(V)3
113-598-2024

Conditions: Performed in a tactical or strategic telecommunications center and may be performed in an NBC environment, given the AN/UGC-74B(V)3 communications terminal, materials for maintaining the AN/UGC-74B(V)3, DA PAM 738-750, TM 11-5815-602-10-1, TM 11-5815-602-24, and DA Form 2404 (Equipment Inspection and Maintenance Worksheet) or DA Form 5988-E (Equipment Inspection Maintenance Worksheet (EGA)).

NOTE: Supervision and assistance are available.

Standards: Performed unit level PMCS, completed all required forms within 15 minutes, and reported all uncorrectable items to the team chief.

Performance Steps

1. Perform PMCS on terminal communications AN/UGC-74B(V)3 as required. (Refer to TM 11-5815-602-10-1 and TM 11-5815-602-24.)

2. Complete DA Form 2404 or DA Form 5988-E. (Refer to DA PAM 738-750.)

3. Report all uncorrectable defects. (Refer to DA PAM 738-750.)

NOTE: Specific extracts of references are not included for this task because of the quantity of material.

Evaluation Preparation: Setup: Provide Soldier with the AN/UGC-74( ), DA PAM 738-750, DA Form 2404 or DA Form 5988-E, TM 11-5815-602-10-1, TM 11-5815-602-24, and materials for maintaining the AN/UGC-74B(V)3.

Brief Soldier: Tell the Soldier to perform unit level PMCS on the AN/UGC-74B(V)3.

Performance Measures	GO	NO GO
1. Performed PMCS on the AN/UGC-74B(V)3 as required.	——	——
2. Completed DA Form 2404 or DA Form 5988-E.	——	——
3. Reported all uncorrectable defects.	——	——

Evaluation Guidance: Score the Soldier a GO if all PMs are passed. Score the Soldier a NO-GO if any PM is failed. If the Soldier fails any PM, show what was done wrong and how to do it correctly. Have the Soldier perform the PMs until they are done correctly.

References

Required	Related
DA FORM 2404	
DA FORM 5988-E	
DA PAM 738-750	
TM 11-5815-602-10-1	
TM 11-5815-602-24	

STP 11-25C13-SM-TG

Operate Digital Message Data Generator, KL-43, or Similar Device
113-609-2214

Conditions: Given a digital message data generator KL-43, appropriate COMSEC key tape, four known good AA batteries, operational radio set, appropriate cables, message, distant station with same equipment, and operator's manual.

Standards: Performed installation and power up, loaded COMSEC, sent an encrypted message to distant station and received an encrypted message from distant station, decrypted the received message IAW with the operator's manual.

Performance Steps

1. Prepare KL-43 for operation (installation).
    a. Ensure equipment is off.
    b. Install four AA batteries.

2. Power up procedures.
    a. Press SRCH (ON) key.
    b. Watch two-line display show Confirm - Turn the unit ON (Y/N).
    c. Press [Y].

3. Load COMSEC.
    a. After power up, the display will show all available FILL positions and all loaded FILL positions. Select one of the available positions by pressing the two (2) numbers (example 01 - 16).
    b. The display shows "Enter the key name." At this time name the key in a way that can be remembered. Press ENTER.
    c. Enter Key Sets from the crypto type.
        (1) Enter KEY SET 1.
        (2) Press ENTER.
        (3) Enter KEY SET 2.
        (4) Press ENTER.
        (5) Enter KEY SET 3.
        (6) Press ENTER.
        (7) Enter KEY SET 4.
        (8) Press ENTER.

4. Prepare a message for transmission.
    a. Pre-mission setup.
        (1) Select the "Set Time and Date" from the Main Menu to ensure that all KL-43s are set within 4 minutes of each other. Press XIT to save settings.
        (2) If Quiet Operations (Q) is desired, it must be enabled at the Main Menu and then press XIT to save setting.

NOTE: A KL-43 connection cable or a fill cable is required for Quiet Operations.

    b. Select "Word Processor" (W) from the main menu.
    c. Select "Message A or B" if a text already exists in the message you selected. The display will show "Do you want to clear message from memory? (Y/N)." You can clear the existing message and start a new one or not clear the message and edit it.
    d. Select "P" to type your message in plain text mode.
    e. Type classification, press ENTER.
    f. Type message and [XIT] when finished.

5. Encrypt message(s).
    a. Select "E" from the Main Menu Encrypt message.
    b. Select the desired message (A or B). If the key is correct, type "Y"; if not, type "N" and the KL-43 will prompt you to select another key.

STP 11-25C13-SM-TG

Performance Steps

        c. After the encryption is complete, the message is ready to send.

   6. Send a Message (after message is typed in and encrypted).
        a. Select (C) "Communication" in the Main Menu.
        b. Select (A) "Audio Data."
        c. Select (A) "Acoustic Coupler" if you are not using a cable, "Connector Audio" (C) if you are using a Fill Cable, or if (Q) "Quiet Operations" is enabled (refer to paragraph C2).
        d. Select (T) "Transmit."

NOTE: If you selected "Acoustic Coupler," the KL-43 will prompt you if you desire to make commo using "U.S. lines or European lines." This is only important if communications are to be made over telephone lines.

        e. Select U.S. lines or European lines (if needed).
        f. Select from the KL-43 prompt the message to send (A or B).
        g. Then prompt you to wait. Press [enter] when ready, or press [XIT] to return to the main menu. The KL-43 will start transmitting as soon as you press [enter].
        h. The KL-43 will display transmission was complete.
        i. Select from the KL-43 prompt either "do you want to transmit again or return to the main menu." Make the appropriate selection.

   7. Receive a Message.
        a. Press "C" (Communications) from the Main Menu.
        b. The KL-43 will prompt you if the Current Default Settings are correct. If correct press "Y"; if not, press "N" and at this time you can make necessary changes.
        c. The KL-43 with either prompts you to transmit or receive, press "R."

NOTE: If the memory is full, you will be prompt if you want to erase one of the messages from memory and which one.

        d. The KL-43 will display "Press enter when ready." Once you press [enter] the display will show "waiting for carrier." At this time you are ready to receive.
        e. Once the message is received, the KL-43 will let you know if it is stored as A or B.

NOTE: Remember where the message is stored.

   8. Decrypt the Received Message.
        a. From the Main Menu select "D" (Decrypt). The KL-43 will prompt you if the crypto is correct.
        b. Select (Y) "Yes" if correct. Change it if it is not.
        c. To review the message press "R" (Review) from the "Main Menu." The KL-43 will prompt you which message. Select the appropriate message.

   9. Zeroize the COMSEC.
        a. Press the ZERO key on the keypad.
        b. The KL-43 will prompt you to enter the key ID of the key to be zeroized (example 01 or "A" for ALL). Then ask you to confirm your choice.

   10. Power Down the KL-43.
        a. From Main Menu, select (O) "Turn Unit Off."
        b. Remove batteries if unit will not be used in the near future.

Performance Measures	**GO**	**NO GO**
1. Prepared the KL-43 for operation (installation).	——	——
2. Performed power up procedures.	——	——
3. Loaded COMSEC.	——	——
4. Prepared a message for transmission.	——	——

STP 11-25C13-SM-TG

Performance Measures	**GO**	**NO GO**
5. Encrypted message(s).	____	____
6. Sent a Message.	____	____
7. Received a Message.	____	____
8. Decrypted the received message.	____	____
9. Zeroized the COMSEC.	____	____
10. Powered down the KL-43.	____	____

Evaluation Guidance: Score the Soldier a GO if all PMs are passed. Score the Soldier a NO-GO if any PM is failed. If the Soldier fails any PM, show what was done wrong and how to do it correctly. Have the Soldier perform the PMs until they are done correctly.

References
    Required   **Related**
    OM

STP 11-25C13-SM-TG

Install Communications Central AN/TSC-128
113-620-1044

Conditions: Given a shelter-mounted communications central AN/TSC-128 installed on a vehicle, ground rod, ground strap, and pliers; 5-pound sledgehammer; 8-inch adjustable wrench and 8-inch flat-tip screwdriver; FM 24-18, TB 43-0129, TM 11-5805-201-12, and TM 11-2300-476-14&P.

Standards: Sited and grounded the communications central set, erected the whip antenna, connected the selected power cable, installed the security equipment, positioned the boarding ladder, and tuned and operated the communications central set within 30 minutes.

Performance Steps
(Refer to TM 11-5805-201-12 for all performance steps.)

1. Check equipment for completeness.
2. Site communications central.
3. Position vehicle boarding ladder.
4. Erect shade tarpaulin.
5. Install COMSEC device in racks and connect signal cables.
6. Perform preoperational procedures.
7. Erect whip antenna.
8. Ground rod installation/ground shelter.
9. Connect appropriate power cable.

Performance Measures (Refer to TM 11-5805-201-12 for all PMs.)	GO	NO GO
1. Checked equipment for completeness.	——	——
2. Selected site for communications central.	——	——
3. Positioned vehicle boarding ladder.	——	——
4. Erected shade tarpaulin.	——	——
5. Installed COMSEC device in racks and connected signal cables.	——	——
6. Performed preoperational procedures.	——	——
7. Erected whip antenna.	——	——
8. Grounded rod installation/grounded shelter.	——	——
9. Connected appropriate power cable.	——	——

Evaluation Guidance: Score the Soldier a GO if all PMs are passed. Score the Soldier a NO-GO if any PM is failed. If the Soldier fails any PM, show what was done wrong and how to do it correctly. Have the Soldier perform the PMs until they are done correctly.

References
- **Required**
  - FM 24-18
  - TB 43-0129
  - TM 11-2300-476-14&P
  - TM 11-5805-201-12
- **Related**
  - TM 11-5815-334-10
  - TM 11-5815-602-10

STP 11-25C13-SM-TG

Operate Communications Central AN/TSC-128
113-620-2049

Conditions: Given an installed operational communications central AN/TSC-128, distant station, a message to transmit, unit SOI, TB 43-0129, TM 11-5805-201-12, TM 11-5820-924-13, TM 11-5815-334-10, TM 11-5815-602-10, and TM 11-2300-475-13&P1.

Standards: Performed all the proper procedures to place the AN/TSC-128 in the selected mode of operation and to shut it down IAW TM 11-5820-924-13 within 30 minutes.

Performance Steps

1. Determine operating frequency from current unit SOI.
2. Perform preliminary starting procedures.
3. Perform operational equipment settings.
4. Perform preoperational equipment checks.
5. Perform operational and self-test of communications terminal AN/UGC-74B(V)3.
6. Perform tuning procedures.
7. Select a precedence message.
8. Transmit a message.

Performance Measures	GO	NO GO
1. Determined operating frequency from current unit SOI.	——	——
2. Performed preliminary starting procedures.	——	——
3. Performed preoperational equipment settings.	——	——
4. Performed preoperational equipment checks.	——	——
5. Performed operational and self-test of communications terminal AN/UGC-74B(V)3.	——	——
6. Performed tuning procedures.	——	——
7. Selected a precedence message.	——	——
8. Transmitted a message.	——	——

Evaluation Guidance: Score the Soldier a GO if all PMs are passed. Score the Soldier a NO-GO if any PM is failed. If the Soldier fails any PM, show what was done wrong and how to do it correctly. Have the Soldier perform the PMs until they are done correctly.

References
    Required   **Related**
    TB 43-0129   SB 11-131-1
    TM 11-2300-475-13&P-1
    TM 11-5805-201-12
    TM 11-5815-334-10
    TM 11-5815-602-10
    TM 11-5820-924-13
    UNIT SOI

STP 11-25C13-SM-TG

Perform Unit Level Maintenance on Communications Central AN/TSC-128
113-620-3064

Conditions: Given communications central AN/TSC-128; 8-inch flat-tip screwdriver, 8-inch adjustable wrench, and replaceable fuses as required; mild detergent solution and cleaning fluid trichloroethane; clean, dry, lint-free cloth; eraser; DA PAM 738-750, TM 11-5820-924-13, TM 11-5805-201-12, TM 11-5815-334-10, TM 11-5815-602-10, TB 43-0129, and DA Form 2404 or DA Form 5988-E.

Standards: Corrected any discovered faults according to PM 2, recorded uncorrectable defects on DA Form 2404 or DA Form 5988-E, and reported to your immediate supervisor IAW DA PAM 738-750.

Performance Steps

1. Perform operator's troubleshooting procedures on communications central AN/TSC-128. (Refer to TM 11-5820-924-13.)

DANGER: Dangerous voltages exist at the CU-2064, 50-OHM LINE, and WHIP antenna connectors. Be careful when working around the antenna or antenna connectors. Radio frequency voltages as high as 10,000 volts exist at these points. Operator and maintenance personnel should be familiar with the requirements of TB 43-0129 before attempting installation or operation of communications central AN/TSC-128. Injury or DEATH could result from improper or careless operation.

2. Correct defects if necessary. Replacement parts or materials are obtained from your team chief.

3. Complete DA Form 2404 or DA Form 5988-E as a daily maintenance form. (Refer to DA PAM 738-750.)

4. Report all uncorrectable defects. (Refer to DA PAM 738-750.)
    a. Notify immediate supervisor of all uncorrectable faults found.
    b. Submit DA Form 2404 or DA Form 5988-E to supervisor or support maintenance personnel.

Performance Measures	GO	NO GO
1. Performed operator's troubleshooting procedures on communications central AN/TSC-128. (Refer to TM 11-5820-924-13.)	——	——
2. Corrected defects if necessary. Replacement parts or materials were obtained from team chief.	——	——
3. Completed DA Form 2404 or DA Form 5988-E as a daily maintenance form. (Refer to DA PAM 738-750.)	——	——
4. Reported all uncorrectable defects. (Refer to DA PAM 738-750.)	——	——

Evaluation Guidance: Score the Soldier a GO if all PMs are passed. Score the Soldier a NO-GO if any PM is failed. If the Soldier fails any PM, show what was done wrong and how to do it correctly. Have the Soldier perform the PMs until they are done correctly.

References
    Required   **Related**
    DA FORM 2404
    DA FORM 5988-E
    DA PAM 738-750
    TB 43-0129
    TM 11-2300-475-13&P-1
    TM 11-5805-201-12
    TM 11-5815-334-10
    TM 11-5815-602-10
    TM 11-5820-924-13
    UNIT SOI

STP 11-25C13-SM-TG

Subject Area 4: GENERATORS

Install Generator Set AN/MJQ-35 5 KW
113-601-1039

Conditions: Given a tactical or nontactical situation, generator set AN/MJQ-35 or MEP-802A, ground rod, ground strap and pliers, 8-inch adjustable wrench and 8-inch flat-tip screwdriver, slide hammer, TM 9-6115-659-13&P, and TM 9-6115-641-10.

NOTE: Supervision and assistance are available.

Standards: Within 30 minutes, sited and grounded the generator, connected power cable, determined fuel supply, and connected without causing damage to any connectors or the generator set, and the generator was ready for operation according to PMs 1 through 5.

Performance Steps

1. Site generator set.

DANGER: DO NOT disconnect trailer from towing vehicle before brakes are set and front landing leg/support leg is lowered. Failure to observe this warning could result in severe personal injury or DEATH from trailer tipping or rolling.

    a. Locate the trailer on as level a surface as possible for efficient operation of the generator set.
    b. Using the two handbrake levers, set the trailer brakes securely to prevent any movement.
    c. Adjust front landing leg using elevation crank to level the trailer.
    d. Turn the base of the rear leveling support jack until it makes firm contact with the ground.

2. Ground generator set.
    a. Secure ground rod to generator ground stud.
    b. Secure ground strap to ground rod.
    c. Drive the ground rod 8 feet or deeper into the ground to provide an effective ground.

3. Connect external fuel line (if required). Place the external fuel source several feet, but no more than 25 feet, away from the generator set.

DANGER: The fuel in this generator set is highly explosive. DO NOT smoke or use open flame when performing maintenance. Flames and explosion could result in severe personal injury or DEATH.

4. Connect power cables for operation under the correct voltage.

DANGER: Ensure generator sets are shut down before connecting load cables. Failure to observe this warning can cause severe personal injury or DEATH.

5. Assure AC Voltage Reconnection switch is set to the desired voltage/phase position.

Evaluation Preparation: Setup: Ensure equipment is complete and operational.

Brief Soldier: Tell the Soldier all PMs must be performed in sequence and completed correctly within 20 minutes.

Performance Measures   GO   NO GO

   1. Sited generator set.   ____   ____

   2. Grounded generator set.   ____   ____

   3. Connected external fuel line (if required).   ____   ____

Evaluation Guidance: Score the Soldier a GO if all PMs are passed. Score the Soldier a NO-GO if any PM is failed. If the Soldier fails any PM, show what was done wrong and how to do it correctly. Have the Soldier perform the PMs until they are done correctly.

References
   Required   **Related**
   TM 9-6115-641-10
   TM 9-6115-659-13&P

Install Generator Set AN/MJQ-37 10 KW
113-601-1040

Conditions: Given a tactical or nontactical situation, generator set AN/MJQ-37, ground rod, ground strap and pliers, 8-inch adjustable wrench and 8-inch flat-tip screwdriver, slide hammer, TM 9-6115-642-10, and TM 9-6115-660-13&P.

NOTE: Supervision and assistance are available.

Standards: Within 30 minutes, sited and grounded the generator, connected power cable, determined and connected fuel supply without causing damage to any connectors or the generator set, and the generator was ready for operation according to PMs 1 through 5.

Performance Steps

1. Site generator set.

DANGER: Do not disconnect trailer from towing vehicle before brakes are set and front landing leg/support leg is lowered. Failure to observe this warning could result in severe personal injury or DEATH from trailer tipping or rolling.

    a. Locate the trailer on as level a surface as possible for efficient operation of the generator set.
    b. Using the two handbrake levers, set the trailer brakes securely to prevent any movement.
    c. Adjust front landing leg using elevation crank to level the trailer.
    d. Turn the base of the rear leveling support jack until it makes firm contact with the ground.

2. Ground generator set.
    a. Secure ground rod to generator ground stud.
    b. Secure ground strap to ground rod.
    c. Drive the ground rod 8 feet or deeper into the ground to provide an effective ground.

3. Connect external fuel line (if required). Place the external fuel source several feet, but no more than 25 feet away from the generator set.

DANGER: The fuel in this generator set is highly explosive. DO NOT smoke or use open flame when performing maintenance. Flames and explosion could result in severe personal injury or DEATH.

4. Connect power cables for operation under the correct voltage.

DANGER: Ensure generator sets are shut down before connecting load cables. Failure to observe this warning can cause severe personal injury or death.

5. Assure AC Voltage Reconnection switch is set to the desired voltage/phase position.

Evaluation Preparation: Setup: Ensure equipment is complete and operational.

Brief Soldier: Tell the Soldier all PMs must be performed in sequence and completed correctly within 20 minutes.

Performance Measures	GO	NO GO
1. Properly sited generator set.	——	——
2. Grounded generator set.	——	——
3. Connected external fuel line (if required).	——	——
4. Connected power cables for operation under the correct voltage.	——	——
5. Assured AC Voltage Reconnection switch was set to the desired voltage/phase position.	——	——

Evaluation Guidance: Score the Soldier a GO if all PMs are passed. Score the Soldier a NO-GO if any PM is failed. If the Soldier fails any PM, show what was done wrong and how to do it correctly. Have the Soldier perform the PMs until they are done correctly.

References
    Required   **Related**
    TM 9-6115-642-10
    TM 9-6115-660-13&P

STP 11-25C13-SM-TG

Install Generator Set AN/MJQ-42 3 KW
113-601-1041

Conditions: As a radio operator-maintainer in a field environment, given a tactical or nontactical situation, generator set AN/MJQ-42 or MEP-701A, ground rod, ground strap and pliers, 8-inch adjustable wrench and 8-inch flat-tip screwdriver, slide hammer, TM 5-6115-640-14&P, TM 9-2330-202-14&P, and TM 5-6115-615-12.

NOTE: Supervision and assistance are available.

Standards: Within 30 minutes, sited and grounded the generator, connected power cable, determined and connected fuel supply without causing damage to any connectors or the generator set, and the generator was ready for operation according to PMs 1 through 5.

Performance Steps

1. Site generator set.

DANGER: DO NOT disconnect trailer from towing vehicle before brakes are set and front landing leg/support leg is lowered. Failure to observe this warning could result in severe personal injury or DEATH from trailer tipping or rolling.

    a. Locate the trailer on as level a surface as possible for efficient operation of the generator set.
    b. Using the two handbrake levers, set the trailer brakes securely to prevent any movement.
    c. Adjust front landing leg using elevation crank to level the trailer.
    d. Turn the base of the rear leveling support jack until it makes firm contact with the ground.

2. Ground generator set.
    a. Secure ground rod to generator ground stud.
    b. Secure ground strap to ground rod.
    c. Drive the ground rod 8 feet or deeper into the ground to provide an effective ground.

3. Connect external fuel line (if required). Place the external fuel source several feet, but no more than 25 feet away from the generator set.

DANGER: The fuel in this generator set is highly explosive. DO NOT smoke or use open flame when performing maintenance. Flames and explosion could result in severe personal injury or DEATH.

4. Connect power cables for operation under the correct voltage.

DANGER: Ensure generator sets are shut down before connecting load cables. Failure to observe this warning can cause severe personal injury or DEATH.

5. Assure AC Voltage Reconnection switch is set to the desired voltage/phase position.

Evaluation Preparation: Setup: Ensure equipment is complete and operational.

Brief Soldier: Tell the Soldier all PMs must be performed in sequence and completed correctly within 20 minutes.

Performance Measures	**GO**	**NO GO**
1. Properly sited generator set.	——	——
2. Grounded generator set.	——	——
3. Connected external fuel line (if required).	——	——
4. Connected power cables for operation under the correct voltage.	——	——
5. Assured AC Voltage Reconnection switch was set to the desired voltage/phase position.	——	——

Evaluation Guidance: Score the Soldier a GO if all PMs are passed. Score the Soldier a NO-GO if any PM is failed. If the Soldier fails any PM, show what was done wrong and how to do it correctly. Have the Soldier perform the PMs until they are done correctly.

References

Required	**Related**
TM 5-6115-615-12	
TM 5-6115-640-14&P	
TM 9-2330-202-14&P	

STP 11-25C13-SM-TG

Operate Generator Set AN/MJQ-35 5 KW
113-601-2054

Conditions: Given a tactical or nontactical situation, generator set AN/MJQ-35 or MEP-802A, TM 9-6115-659-13&P, and TM 9-6115-641-10.

NOTE: Supervision and assistance are available.

Standards: Within 30 minutes placed the generator set into operation and took out of operation IAW PMs 1 through 3 and PM 14.

NOTE: PMs 4 through 13 are evaluated as necessary.

Performance Steps

1. Perform starting procedures. (Refer to TM 9-6115-641-10.)

DANGER: Never attempt to start the generator set if it has not been properly grounded. Failure to observe this warning could result in serious injury or DEATH by electrocution.

WARNING: Do not crank engine in excess of 15 seconds. Allow starter to cool at least 15 seconds between attempted starts. Failure to observe this caution could result in damage to the starter.

2. Start generator set. (Refer to TM 9-6115-641-10.)

3. Perform operating procedures. (Refer to TM 9-6115-641-10 and TM 9-6115-659-13&P.)

DANGER: High voltage is produced when generator set is in operation. Improper operation could result in personal injury or DEATH by electrocution.

NOTE: Under normal conditions warn up engine without load for five minutes. (If required, load can be applied immediately.)

CAUTION: With any access door open, the noise level of this generator set when operating could cause hearing damage. Hearing protection must be worn when working near the generator set while running.

4. Operation in extreme cold weather below -25 degrees F (-31 degrees C). (Refer to TM 9-6115-641-10.)

5. Operation in extreme heat above 120 degrees F (49 degree C). (Refer to TM 9-6115-641-10.)

6. Operation in dusty or sandy areas. (Refer to TM 9-6115-641-10.)

7. Operation under rainy or humid conditions. (Refer to TM 9-6115-641-10.)

8. Operation in salt-water areas. (Refer to TM 9-6115-641-10.)

9. Operation at high altitudes. (Refer to TM 9-6115-641-10.)

10. NATO slave receptacle start operation. (Refer to TM 9-6115-641-10.)

11. Emergency stopping. (Refer to TM 9-6115-641-10.)

12. Operation using Battle Short switch. (Refer to TM 9-6115-641-10.)

13. Operation while in contaminated areas. (Refer to TM 9-6115-641-10.)

14. Perform stopping procedures. (Refer to TM 9-6115-641-10 and TM 9-6115-659-13&P.)

STP 11-25C13-SM-TG

Evaluation Preparation: Setup: Ensure equipment is complete and operational. Select appropriate PMs 4 through 13.

Brief Soldier: Tell the Soldier he must perform in sequence and pass PMs 1 through 3 and PM 14. In addition, the supervisor will select appropriate PMs 4 through 13, which must be completed within 20 minutes.

Performance Measures	GO	NO GO
1. Performed starting procedures. (Refer to TM 9-6115-641-10.)	——	——
2. Started generator set. (Refer to TM 9-6115-641-10.)	——	——
3. Performed operating procedures. (Refer to TM 9-6115-641-10 and TM 9-6115-659-13&P.)	——	——
4. Operation in extreme cold weather below -25 degrees F (-31 degrees C). (Refer to TM 9-6115-641-10.)	——	——
5. Operation in extreme heat above 120 degrees F (49 degree C). (Refer to TM 9-6115-641-10.)	——	——
6. Operation in dusty or sandy areas. (Refer to TM 9-6115-641-10.)	——	——
7. Operation under rainy or humid conditions. (Refer to TM 9-6115-641-10.)	——	——
8. Operation in salt-water areas. (Refer to TM 9-6115-641-10.)	——	——
9. Operation at high altitudes. (Refer to TM 9-6115-641-10.)	——	——
10. NATO slave receptacle start operation. (Refer to TM 9-6115-641-10.)	——	——
11. Emergency stopping. (Refer to TM 9-6115-641-10.)	——	——
12. Operation using Battle Short switch. (Refer to TM 9-6115-641-10.)	——	——
13. Operation while in contaminated areas. (Refer to TM 9-6115-641-10.)	——	——
14. Performed stopping procedures. (Refer to TM 9-6115-641-10 and TM 9-6115-659-13&P.)	——	——

Evaluation Guidance: Score the Soldier a GO if all PMs are passed. Score the Soldier a NO-GO if any PM is failed. If the Soldier fails any PM, show what was done wrong and how to do it correctly. Have the Soldier perform the PMs until they are done correctly.

References
    Required   **Related**
    TM 9-6115-641-10
    TM 9-6115-659-13&P

STP 11-25C13-SM-TG

Perform Operator's Troubleshooting Procedures on Generator Set AN/MJQ-35 5 KW
113-601-2055

Conditions: Given a tactical or nontactical situation, generator set AN/MJQ-35 or MEP-802A, fuses as required, one quart of oil and clean rags, 8-inch adjustable wrench and 8-inch flat-tip screwdriver, DA PAM 738-750, TM 9-6115-659-13&P, TM 9-6115-641-10, and DA Form 2404 or DA Form 5988-E.

NOTE: Supervision and assistance are available.

Standards: Corrected any discovered faults IAW TM 9-6115-641-10 and recorded uncorrectable faults on DA Form 2404 or DA Form 5988-E, without error, and reported to your immediate supervisor according to PMs 1 through 4.

Performance Steps

1. Perform operator's troubleshooting procedures on generator set AN/MJQ-35 5 KW. (Refer to TM 9-6115-641-10.)

2. Correct defects if necessary. Replacement parts or materials are obtained from your team chief. (Refer to TM 9-6115-641-10.)

3. Complete DA Form 2404 or DA Form 5988-E as a daily maintenance form. (Refer to DA PAM 738-750.)

4. Report all uncorrectable defects. (Refer to DA PAM 738-750.)

Evaluation Preparation: Setup: Ensure equipment has preplanned defects.

Brief Soldier: Tell the Soldier all PMs must be passed in sequence.

Performance Measures	GO	NO GO
1. Performed operator's troubleshooting procedures on generator set AN/MJQ-35 5 KW. (Refer to TM 9-6115-641-10.)	——	——
2. Corrected defects when necessary. Replacement parts or materials were obtained from your team chief. (Refer to TM 9-6115-641-10.)	——	——
3. Completed DA Form 2404 or DA Form 5988-E as a daily maintenance form. (Refer to DA PAM 738-750.)	——	——
4. Reported all uncorrectable defects. (Refer to DA PAM 738-750.)	——	——

Evaluation Guidance: Score the Soldier a GO if all PMs are passed. Score the Soldier a NO-GO if any PM is failed. If the Soldier fails any PM, show what was done wrong and how to do it correctly. Have the Soldier perform the PMs until they are done correctly.

References

Required	Related
DA FORM 2404	
DA FORM 5988-E	
DA PAM 738-750	
TM 9-6115-641-10	
TM 9-6115-659-13&P	

STP 11-25C13-SM-TG

Operate Generator Set AN/MJQ-37 10 KW
113-601-2057

Conditions: As a radio operator-maintainer in a field environment, given a tactical or nontactical situation, generator set AN/MJQ-37or MEP-803A, TM 9-6115-642-10, and TM 9-6115-660-13&P.

NOTE: Supervision and assistance are available.

Standards: Within 30 minutes, placed the generator set into and taken out of operation IAW PMs 1 through 3 and PM 14.

NOTE: PMs 4 through 13 are evaluated as necessary.

Performance Steps

1. Perform starting procedures. (Refer to TM 9-6115-642-10.)

DANGER: Never attempt to start the generator set if it has not been properly grounded. Failure to observe this warning could result in serious injury or DEATH by electrocution.

WARNING: Do not crank engine in excess of 15 seconds. Allow starter to cool at least 15 seconds between attempted starts. Failure to observe this caution could result in damage to the starter.

2. Start generator set. (Refer to TM 9-6115-642-10.)

3. Perform operating procedures. (Refer to TM 9-6115-642-10 and TM 9-6115-660-13&P.)

DANGER: High voltage is produced when generator set is in operation. Improper operation could result in personal injury or DEATH by electrocution.

NOTE: Under normal conditions warn up engine without load for 5 minutes. (If required, load can be applied immediately.)

CAUTION: With any access door open, the noise level of this generator set when operating could cause hearing damage. Hearing protection must be worn when working near the generator set while running.

4. Operation in extreme cold weather below -25 degrees F (-31 degrees C). (Refer to TM 9-6115-642-10.)

5. Operation in extreme heat above 120 degrees F (49 degree C). (Refer to TM 9-6115-642-10.)

6. Operation in dusty or sandy areas. (Refer to TM 9-6115-642-10.)

7. Operation under rainy or humid conditions. (Refer to TM 9-6115-642-10.)

8. Operation in salt-water areas. (Refer to TM 9-6115-642-10.)

9. Operation at high altitudes. (Refer to TM 9-6115-642-10.)

10. NATO slave receptacle start operation. (Refer to TM 9-6115-642-10.)

11. Emergency stopping. (Refer to TM 9-6115-642-10.)

12. Operation using Battle Short switch. (Refer to TM 9-6115-642-10.)

13. Operation while in contaminated areas. (Refer to TM 9-6115-642-10.)

14. Perform stopping procedures. (Refer to TM 9-6115-642-10 and TM 9-6115-660-13&P.)

STP 11-25C13-SM-TG

Evaluation Preparation: Setup: Ensure equipment is complete and operational. Select appropriate PMs 4 through 13.

Brief Soldier: Tell the Soldier he must pass PMs 1 through 3 and PM 14; these tasks must be completed in sequence. In addition, the supervisor will select appropriate PMs 4 through 13, which must be completed within 20 minutes.

Performance Measures	**GO**	**NO GO**
1. Performed starting procedures.	——	——
2. Started generator set.	——	——
3. Performed operating procedures.	——	——
4. Operation in extreme cold weather below -25 degrees F (-31 degrees C).	——	——
5. Operation in extreme heat above 120 degrees F (49 degree C).	——	——
6. Operation in dusty or sandy areas.	——	——
7. Operation under rainy or humid conditions.	——	——
8. Operation in salt-water areas.	——	——
9. Operation at high altitudes.	——	——
10. NATO slave receptacle start operation.	——	——
11. Emergency stopping.	——	——
12. Operation using Battle Short switch.	——	——
13. Operation while in contaminated areas.	——	——
14. Performed stopping procedures.	——	——

Evaluation Guidance: Score the Soldier a GO if all PMs are passed. Score the Soldier a NO-GO if any PM is failed. If the Soldier fails any PM, show what was done wrong and how to do it correctly. Have the Soldier perform the PMs until they are done correctly.

References
    Required   **Related**
    TM 9-6115-642-10
    TM 9-6115-660-13&P

STP 11-25C13-SM-TG

Perform Operator's Troubleshooting Procedures and Generator Set AN/MJQ-37 10 KW
113-601-2058

Conditions: As a radio operator-maintainer in a field environment, given a tactical or nontactical situation, generator set AN/MJQ-37 or MEP-803A, fuses as required, one quart of oil and clean rags, 8-inch adjustable wrench and 8-inch flat-tip screwdriver, DA PAM 738-750, TM 9-6115-642-10, TM 9-6115-660-13&P, and DA Form 2404 or DA Form 5988-E.

NOTE: Supervision and assistance are available.

Standards: Corrected any discovered faults IAW TM 9-6115-642-10 and recorded uncorrectable faults on DA Form 2404 or DA Form 5988-E, without error, and reported to your immediate supervisor according to PMs 1 through 4.

Performance Steps

1. Perform operator's troubleshooting procedures on generator set AN/MJQ-37 10 KW. (Refer to TM 9-6115-642-10.)

2. Correct defects if necessary. Replacement parts or materials are obtained from your team chief. (Refer to TM 9-6115-642-10.)

3. Complete DA Form 2404 or DA Form 5988-E as a daily maintenance form. (Refer to DA PAM 738-750.)

4. Report all uncorrectable defects. (Refer to DA PAM 738-750.)

Evaluation Preparation: Setup: Ensure equipment has preplanned defects.

Brief Soldier: Tell the Soldier all PMs must be passed in sequence.

Performance Measures	GO	NO GO
1. Performed operator's troubleshooting procedures on generator set AN/MJQ-37.	——	——
2. Corrected defects if necessary. Replacement parts or materials were obtained from your team chief.	——	——
3. Completed DA Form 2404 or DA Form 5988-E as a daily maintenance form.	——	——
4. Reported all uncorrectable defects.	——	——

Evaluation Guidance: Score the Soldier a GO if all PMs are passed. Score the Soldier a NO-GO if any PM is failed. If the Soldier fails any PM, show what was done wrong and how to do it correctly. Have the Soldier perform the PMs until they are done correctly.

References

Required	Related
DA FORM 2404	
DA FORM 5988-E	
DA PAM 738-750	
TM 9-6115-642-10	
TM 9-6115-660-13&P	

STP 11-25C13-SM-TG

Operate Generator Set AN/MJQ-42 3 KW
113-601-2059

Conditions: As a radio operator-maintainer in a field environment, given a tactical or nontactical situation, generator set AN/MJQ-42 or MEP-701A, TM 5-6115-615-12, TM 9-2330-202-14&P, and TM 5-6115-640-14&P.

NOTE: Supervision and assistance are available.

Standards: Within 30 minutes, placed the generator set into and taken out of operation IAW PMs 1 through 3 and PM 14.

NOTE: PMs 4 through 13 are evaluated as necessary.

Performance Steps

 1. Perform starting procedures. (Refer to TM 5-6115-615-12.)

DANGER: Never attempt to start the generator set if it has not been properly grounded. Failure to observe this warning could result in serious injury or DEATH by electrocution.

WARNING: Do not crank engine in excess of 15 seconds. Allow starter to cool at least 15 seconds between attempted starts. Failure to observe this caution could result in damage to the starter.

 2. Start generator set. (Refer to TM 5-6115-615-12.)

 3. Perform operating procedures. (Refer to TM 5-6115-615-12 and TM 5-6115-640-14&P.)

DANGER: High voltage is produced when generator set is in operation. Improper operation could result in personal injury or DEATH by electrocution.

NOTE: Under normal conditions warn up engine without load for 5 minutes. (If required, load can be applied immediately.)

CAUTION: With any access door open, the noise level of this generator set when operating could cause hearing damage. Hearing protection must be worn when working near the generator set while running.

 4. Operation in extreme cold weather below -25 degrees F (-31 degrees C). (Refer to TM 5-6115-615-12.)

 5. Operation in extreme heat above 120 degrees F (49 degree C). (Refer to TM 5-6115-615-12.)

 6. Operation in dusty or sandy areas. (Refer to TM 5-6115-615-12.)

 7. Operation under rainy or humid conditions. (Refer to TM 5-6115-615-12.)

 8. Operation in salt-water areas. (Refer to TM 5-6115-615-12.)

 9. Operation at high altitudes. (Refer to TM 5-6115-615-12.)

 10. NATO slave receptacle start operation. (Refer to TM 5-6115-615-12.)

 11. Emergency stopping. (Refer to TM 5-6115-615-12.)

 12. Operation using Battle Short switch. (Refer to TM 5-6115-615-15.)

 13. Operation while in contaminated areas. (Refer to TM 5-6115-615-12.)

 14. Perform stopping procedures. (Refer to TM 5-6115-615-12 and TM 5-6115-640-14&P.)

STP 11-25C13-SM-TG

Evaluation Preparation: Setup: Ensure equipment is complete and operational. Select appropriate PMs 4 through 13.

Brief Soldier: Tell the Soldier he must pass PMs 1 through 3 and PM 14; these tasks must be completed in sequence. In addition, the supervisor will select appropriate PMs 4 through 13, which must be completed within 20 minutes.

Performance Measures	**GO**	**NO GO**
1. Performed starting procedures.	——	——
2. Started generator set.	——	——
3. Performed operating procedures.	——	——
4. Operation in extreme cold weather below -25 degrees F (-31 degrees C).	——	——
5. Operation in extreme heat above 120 degrees F (49 degree C).	——	——
6. Operation in dusty or sandy areas.	——	——
7. Operation under rainy or humid conditions.	——	——
8. Operation in salt-water areas.	——	——
9. Operation at high altitudes.	——	——
10. NATO slave receptacle start operation.	——	——
11. Emergency stopping.	——	——
12. Operation using Battle Short switch.	——	——
13. Operation while in contaminated areas.	——	——
14. Performed stopping procedures.	——	——

Evaluation Guidance: Score the Soldier a GO if all PMs are passed. Score the Soldier a NO-GO if any PM is failed. If the Soldier fails any PM, show what was done wrong and how to do it correctly. Have the Soldier perform the PMs until they are done correctly.

References
    Required   **Related**
    TM 5-6115-615-12
    TM 5-6115-640-14&P
    TM 9-2330-202-14&P

STP 11-25C13-SM-TG

Perform Operator's Troubleshooting Procedures on Generator Set AN/MJQ-42 3 KW
113-601-2060

Conditions: As a radio operator-maintainer in a field environment, given a tactical or nontactical situation, generator set AN/MJQ-42 or MEP-701A, fuses as required, one quart of oil and clean rags, 8-inch adjustable wrench and 8-inch flat-tip screwdriver, DA PAM 738-750, TM 5-6115-640-14&P, TM 9-2330-202-14&P, TM 5-6115-615-12, and DA Form 2404 or DA Form 5988-E.

NOTE: Supervision and assistance are available.

Standards: Corrected any discovered faults IAW TM 9-6115-641-10 and recorded uncorrectable faults on DA Form 2404 or DA Form 5988-E, without error, and reported to your immediate supervisor according to PMs 1 through 4.

Performance Steps

1. Perform operator's troubleshooting procedures on generator set AN/MJQ-42 (Refer to TM 5-6115-640-14&P.)

2. Correct defects if necessary. Replacement parts or materials are obtained from your team chief. (Refer to TM 9-6115-640-14&P.)

3. Complete DA Form 2404 or DA Form 5988-E as a daily maintenance form. (Refer to DA PAM 738-750.)

4. Report all uncorrectable defects. (Refer to DA PAM 738-750.)

Evaluation Preparation: Setup: Ensure equipment has preplanned defects.

Brief Soldier: Tell the Soldier all PMs must be passed in sequence.

Performance Measures	GO	NO GO
1. Performed operator's troubleshooting procedures on generator set AN/MJQ-42.	——	——
2. Corrected defects if necessary. Replacement parts or materials were obtained from your team chief.	——	——
3. Completed DA Form 2404 or DA Form 5988-E as a daily maintenance form.	——	——
4. Reported all uncorrectable defects.	——	——

Evaluation Guidance: Score the Soldier a GO if all PMs are passed. Score the Soldier a NO-GO if any PM is failed. If the Soldier fails any PM, show what was done wrong and how to do it correctly. Have the Soldier perform the PMs until they are done correctly.

References

Required	Related
DA FORM 2404	
DA FORM 5988-E	
DA PAM 738-750	
TM 5-6115-615-12	
TM 5-6115-640-14&P	
TM 9-2330-202-14&P	

STP 11-25C13-SM-TG

Perform Unit Level Maintenance on Generator Sets 3 KW - 10 KW
113-601-3016

Conditions: Given a tactical situation, under all weather conditions; generator set 5 KW, organizational vehicle maintenance (OVM) tool; cleaning materials; DA PAM 738-750, TM 5-6115-332-14, DA Form 2404 or DA Form 5988-E, and replacement parts as needed. This task may be performed in an NBC environment.

Standards: Did a complete check of unit level maintenance and recorded all uncorrectable faults on DA Form 2404 or DA Form 5988-E, without error, and reported to the immediate supervisor. Due to the nature of this task, a time limit is not established.

Performance Steps

NOTE: The Soldier must pass all performance steps in sequence.

1. Perform daily PMCS on the generator set. (Refer to TM 5-6115-332-14.)

2. Correct faults if necessary. (Refer to TM 5-6115-332-14.)

3. Complete DA Form 2404 or DA Form 5988-E. (Refer to DA PAM 738-750.)

4. Report all uncorrectable faults. (Refer to DA PAM 738-750.)
    a. Notify immediate supervisor of all uncorrectable faults.
    b. Submit DA Form 2404 or DA Form 5988-E to supervisor or support maintenance personnel.

NOTE: Specific extracts of references are not included for this task because of the quantity of material.

Evaluation Preparation: Setup: Provide the Soldier with a generator set, appropriate OVM, TMs, DA PAM 738-750, and DA Form 2404 or DA Form 5988-E.

Brief Soldier: Tell the Soldier to perform unit level maintenance on the generator set. List all uncorrectable faults on DA Form 2404 or DA Form 5988-E.

Performance Measures	GO	NO GO
NOTE: The Soldier must pass all PMs in sequence.		
1. Performed daily PMCS on the generator set. (Refer to TM 5-6115-332-14.)	——	——
2. Corrected faults when necessary. (Refer to TM 5-6115-332-14.)	——	——
3. Completed DA Form 2404 or DA Form 5988-E. (Refer to DA PAM 738-750.)	——	——
4. Reported all uncorrectable faults. (Refer to DA PAM 738-750.)	——	——

Evaluation Guidance: Score the Soldier a GO if all PMs are passed. Score the Soldier a NO-GO if any PM is failed. If the Soldier fails any PM, show what was done wrong and how to do it correctly. Have the Soldier perform the PMs until they are done correctly.

References

Required	Related
DA FORM 2404	
DA FORM 5988-E	
DA PAM 738-750	
TM 5-6115-332-14	

STP 11-25C13-SM-TG

Perform PMCS on Generator Set AN/MJQ-35 5 KW
113-601-3030

Conditions: Given a tactical or nontactical situation; a requirement; generator set AN/MJQ-35 5 KW; pliers; one quart of oil; clean, dry, lint-free cloth and mild detergent solution; 8-inch adjustable wrench and 8-inch flat-tip screwdriver, DA PAM 738-750, TM 9-6115-659-13&P, TM 9-6115-641-10, and DA Form 2404 or DA Form 5988-E.

NOTE: Supervision and assistance are available.

Standards: Cleaned the exterior of the generator and tightened all dials and knobs; all cables were in good condition (no cracks or broken connectors); and recorded all uncorrectable faults on DA Form 2404 or DA Form 5988-E, without error, and reported to your immediate supervisor according to PMs 1 through 4.

Performance Steps

1. Perform daily PMCS on generator set AN/MJQ-35 5 KW. (Refer to TM 9-6115-641-10.)

2. Correct defects if necessary. Replacement parts or materials are obtained from your team chief. (Refer to TM 9-6115-641-10.)

3. Complete DA Form 2404 or DA Form 5988-E as a daily maintenance form. (Refer to DA PAM 738-750.)

4. Report all uncorrectable defects. (Refer to DA PAM 738-750.)
   a. Notify immediate supervisor of all uncorrectable faults found.
   b. Submit DA Form 2404 or DA Form 5988-E to supervisor or support maintenance personnel.

Evaluation Preparation: Setup: Ensure equipment has preplanned defects.

Brief Soldier: Tell the Soldier all PMs must be passed in sequence.

Performance Measures	GO	NO GO
1. Performed daily PMCS on generator set AN/MJQ-35 5 KW. (Refer to TM 9-6115-641-10.)	——	——
2. Corrected defects when necessary. Replacement parts or materials were obtained from your team chief. (Refer to TM 9-6115-641-10.)	——	——
3. Completed DA Form 2404 or DA Form 5988-E as a daily maintenance form. (Refer to DA PAM 738-750.)	——	——
4. Reported all uncorrectable defects. (Refer to DA PAM 738-750.)	——	——

Evaluation Guidance: Score the Soldier a GO if all PMs are passed. Score the Soldier a NO-GO if any PM is failed. If the Soldier fails any PM, show what was done wrong and how to do it correctly. Have the Soldier perform the PMs until they are done correctly.

References
    Required　　　　　　　　　　　　**Related**
    DA FORM 2404
    DA FORM 5988-E
    DA PAM 738-750
    TM 9-6115-641-10
    TM 9-6115-659-13&P

STP 11-25C13-SM-TG

Perform PMCS on Generator Set AN/MJQ-37 10 KW
113-601-3031

Conditions: As a radio operator-maintainer in a field environment, given a tactical or nontactical situation; a requirement; generator set AN/MJQ-37 or MEP-803A; pliers; one quart of oil; clean, dry, lint-free cloth and mild detergent solution; 8-inch adjustable wrench and 8-inch flat-tip screwdriver, DA PAM 738-750, TM 9-6115-642-10, TM 9-6115-660-13&P, and DA Form 2404 or DA Form 5988-E.

NOTE: Supervision and assistance are available.

Standards: Cleaned the exterior of the generator and tightened all dials and knobs; all cables were in good condition (no cracks or broken connectors); and recorded all uncorrectable faults on DA Form 2404 or DA Form 5988-E, without error, and reported to your immediate supervisor according to PMs 1 through 4.

Performance Steps

1. Perform daily PMCS on generator set AN/MJQ-37 10 KW. (Refer to TM 9-6115-642-10)

2. Correct defects if necessary. Replacement parts or materials are obtained from your team chief. (Refer to TM 9-6115-642-10)

3. Complete DA Form 2404 or DA Form 5988-E as a daily maintenance form. (Refer to DA PAM 738-750.)

4. Report all uncorrectable defects. (Refer to DA PAM 738-750.)
    a. Notify immediate supervisor of all uncorrectable faults found.
    b. Submit DA Form 2404 or DA Form 5988-E to supervisor or support maintenance personnel.

Evaluation Preparation: Setup: Ensure equipment has preplanned defects.

Brief Soldier: Tell the Soldier all PMs must be passed in sequence.

Performance Measures	GO	NO GO
1. Performed daily PMCS on generator set AN/MJQ-37 10 KW.	——	——
2. Corrected defects when necessary. Replacement parts or materials were obtained from your team chief.	——	——
3. Completed DA Form 2404 or DA Form 5988-E as a daily maintenance form.	——	——
4. Reported all uncorrectable defects.	——	——

Evaluation Guidance: Score the Soldier a GO if all PMs are passed. Score the Soldier a NO-GO if any PM is failed. If the Soldier fails any PM, show what was done wrong and how to do it correctly. Have the Soldier perform the PMs until they are done correctly.

References

Required	Related
DA FORM 2404	
DA FORM 5988-E	
DA PAM 738-750	
TM 9-6115-642-10	
TM 9-6115-660-13&P	

STP 11-25C13-SM-TG

Perform PMCS on Generator Set AN/MJQ-42 3 KW
113-601-3032

Conditions: As a radio operator-maintainer in a field environment, given a tactical or nontactical situation; a requirement; generator set AN/MJQ-42; pliers; one quart of oil; clean, dry, lint-free cloth and mild detergent solution; 8-inch adjustable wrench and 8-inch flat-tip screwdriver, DA PAM 738-750, TM 5-6115-640-14&P, TM 9-2330-202-14&P, and DA Form 2404 or DA Form 5988-E.

NOTE: Supervision and assistance are available.

Standards: Cleaned the exterior of the generator and tightened all dials and knobs; all cables were in good condition (no cracks or broken connectors); and recorded all uncorrectable faults on DA Form 2404 or DA Form 5988-E, without error, and reported to your immediate supervisor according to PMs 1 through 4.

Performance Steps

1. Perform daily PMCS on generator set AN/MJQ-42. (Refer to TM 5-6115-640-14&P.)

2. Correct defects if necessary. Replacement parts or materials are obtained from your team chief. (Refer to TM 5-6115-640-14&P.)

3. Complete DA Form 2404 or DA Form 5988-E as a daily maintenance form.

4. Report all uncorrectable defects. (Refer to DA PAM 738-750.)
   a. Notify immediate supervisor of all uncorrectable faults found.
   b. Submit DA Form 2404 or DA Form 5988-E to supervisor or support maintenance personnel.

Evaluation Preparation: Setup: Ensure equipment has preplanned defects.

Brief Soldier: Tell the Soldier all PMs must be passed in sequence.

Performance Measures	GO	NO GO
1. Performed daily PMCS on generator set AN/MJQ-42.	——	——
2. Corrected defects if necessary. Replacement parts or materials were obtained from your team chief.	——	——
3. Completed DA Form 2404 or DA Form 5988-E as a daily maintenance form.	——	——
4. Reported all uncorrectable defects.	——	——

Evaluation Guidance: Score the Soldier a GO if all PMs are passed. Score the Soldier a NO-GO if any PM is failed. If the Soldier fails any PM, show what was done wrong and how to do it correctly. Have the Soldier perform the PMs until they are done correctly.

References

Required	Related
DA FORM 2404	
DA FORM 5988-E	
DA PAM 738-750	
TM 5-6115-640-14&P	
TM 9-2330-202-14&P	

Subject Area 5: HIGH FREQUENCY (HF) RADIOS

Perform Unit Level Troubleshooting on AN/PRC-150 Radio Set
113-620-0108

Conditions: Given a preprogrammed AN/PRC-150 radio set and Publication Number 10515-0103-4100.

Standards: The radio set was operational and could communicate in a net.

Performance Steps

1. Perform preventive maintenance checks and services. (Refer to Publication Number 10515-0103-4011, Chapter 5, paras 5.1 and 5.2.)

2. Perform the radio start-up procedures. If faults occur, go to performance step 3.
    a. Rotate the function switch to the PT position

NOTE: This initializes the radio software and performs a power on self-test. When this test is complete, the normal operational menu is displayed.

3. Perform initial checks. (Refer to Publication Number 10515-0103-4011, Chapter 3, para 3.5.2.) If fault occurs, go to performance step 4.

4. Identify the fault symptom by initiating a BITE test. (Refer to Publication Number 10515-0103-4011, Chapter 3, para 3.3.1.1.)
    a. Select ALL and press [ENT]. When R/T displays a fault message, go to performance step 5.

5. When the fault code is identified, report the fault to a Level III Maintainer. (Refer to Publication Number 10515-0103-4011, Chapter 5, para 5.2.1.)

6. Verify the fault is fixed.

7. Repeat performance steps 3 through 5. If no other faults are found, go to performance step 8.

8. Press the clear [CLR] button three times on the R/T to return to the preset screen.

9. Resume operations. (Refer to Publication Number 10515-0103-4011, Chapter 3, para 3.6.)

Performance Measures	GO	NO GO
1. Performed preventive maintenance checks and services.	——	——
2. Performed the radio start-up procedures.	——	——
3. Performed initial checks.	——	——
4. Identified a fault symptom by initiating a BITE test.	——	——
5. Identified fault code and reported it to a Level III Maintainer.	——	——
6. Verified the fault was fixed.	——	——
7. Repeated PMs 3 through 5 for each fault.	——	——
8. Pressed the clear [CLR] button to return to preset screen.	——	——
9. Resumed operations.	——	——

Evaluation Guidance: Score the Soldier a GO if all PMs are passed. Score the Soldier a NO-GO if any PM is failed. If the Soldier fails any PM, show what was done wrong and how to do it correctly. Have the Soldier perform the PMs until they are done correctly.

References
Required	**Related**
PUB # 10515-0103-4100	

STP 11-25C13-SM-TG

Install Radio Set AN/GRC-106(*)
113-620-1001

Conditions: Given a tactical or nontactical situation, a requirement, radio set AN/GRC-106(*), vehicle with mount MT-3140/GRC-106 and mast base AB-652/GR installed, pliers, 8-inch flat-tip screwdriver, and TM 11-5820-520-10.

Standards: Within 20 minutes, installed the radio in its mount without damage to the radio or mount, erected antenna, made all cable connections, and the radio was ready to be tuned and operated according to PMs 1 through 4.

Performance Steps

DANGER: Dangerous voltages exist at the AM-3349/GRC-106 50-OHM LINE and WHIP antenna connectors. Be careful when working around the antenna or antenna connectors. Radio frequency voltages as high as 10,000 volts exist at these points. Operator and maintenance personnel should be familiar with the requirements of TB 43-0129 before attempting installation or operation of radio set AN/GRC-106(*). Injury or DEATH could result from improper or careless operation.

NOTE: Tell the Soldier all steps must be performed correctly to receive a GO. (Sequence is not scored.)

(Refer to TM 11-5820-520-10 for all performance steps.)

1. Position the AN/GRC-106(*) in mount MT-3140/GRC-106 securely. (Refer to para 2-6a in TM.)

2. Connect all cables.
    a. Connect ground cable. (Refer to para 2-6b in TM.)
    b. Connect power cables. Cable Assembly, Special Purpose, Electrical CX-10071/U connects the radio set to the power source. Connect one cable to the power connector on the receiver transmitter, RT-834/GRC-106(A) or RT-662/GRC-106 and connect the second CX-10071/U to the primary power connector on the Amplifier, AM-3349/GRC-106(*). (See diagram on page 2-28 in TM.)
    c. Connect RF cable assembly. Connect RF cable CG-409/G (8 inches in length) to the RCVR ANT connector on the amplifier and to the RECEIVER IN connector on the Rcvr/Xmtr.

3. Connect antenna.
    a. Whip antenna. (Refer to para 2-6c in TM.)
    b. Doublet antenna. Connect doublet antenna lead-into the 50-OHM LINE connector on left front of amplifier.

4. Connect audio accessories.
    a. Connect loudspeaker, LS-166/U to one of the two audio connectors on lower left of amplifier.
    b. Connect microphone, M-29A to one of the two audio connectors on lower left side of amplifier.

NOTE: If International Morse Code is used instead of radiotelephone, then the telegraph key KY-116/U must be connected to one of the audio connectors using cable assembly, CX-1852/U instead of the microphone.

Performance Measures	GO	NO GO

DANGER: Dangerous voltages exist at the AM-3349/GRC-106 50-OHM LINE and WHIP antenna connectors. Be careful when working around the antenna or antenna connectors. Radio frequency voltages as high as 10,000 volts exist at these points. Operator and maintenance personnel should be familiar with the requirements of TB 43-0129 before attempting installation or operation of radio set AN/GRC-106(*). Injury or DEATH could result from improper or careless operation.

NOTE: Tell the Soldier all steps must be performed correctly to receive a GO. (Sequence is not scored.)

(Refer to TM 11-5820-520-10 for all PMs.)

 1. Positioned the AN/GRC-106(*) in mount MT-3140/GRC-106 securely. ____ ____

 2. Connected all cables. ____ ____

 3. Connected audio accessories. ____ ____

 4. Installed whip antenna. ____ ____

Evaluation Guidance: Score the Soldier a GO if all PMs are passed. Score the Soldier a NO-GO if any PM is failed. If the Soldier fails any PM, show what was done wrong and how to do it correctly. Have the Soldier perform the PMs until they are done correctly.

References
    Required   **Related**
    TM 11-5820-520-10

STP 11-25C13-SM-TG

Install Radio Set AN/GRC-193A or Similar Radio Sets
113-620-1028

Conditions: Given an operational radio set AN/GRC-193A, operating frequency, a distant station, and TM 11-5820-924-13.

Standards: Performed all of the proper procedures to install the components of radio set AN/GRC-193A for voice operation IAW TM 11-5820-924-13.

Performance Steps

1. Perform installation of the AM-6879 and the RT-1209.
   a. Loosen the wing nuts on the front of the mounting tray.
   b. Insert the back edge of the amplifier-converter assembly under the rear lip at the left side of the mounting tray.
   c. Position the front fasteners on MT-6232 to hold in and down on the front lip of the amplifier-converter.
   d. Tighten the wing nuts firmly.
   e. Secure the receiver/transmitter on the right side of the mounting tray by repeating steps a through d.

2. Perform installation of the AM-6545 and the CU-2064.
   a. Loosen the wing nuts at the right and left side of the mounting tray.
   b. Set the antenna coupler on the top left of the mounting tray.
   c. Position the fasteners of the mount to hold in and down on the side lip of the antenna coupler.
   d. Tighten the wing nuts properly.
   e. Set the power amplifier on the top right of the mounting tray. Position the side fasteners of the mount to hold in and down on the side lip of the power amplifier.
   f. Tighten the wing nuts firmly.

3. Perform cable connections.

CAUTION: Align all cable connectors before mating and fastening.

   a. Connect power control-RF cable to receptacle J6 and connect the other end to receptacle J1.
   b. Connect the short audio cable to the bottom audio connector and the other end to the bottom audio connector.
   c. Install power control cable-to-cable receptacle J2 and the other end to receptacle J1.
   d. Connect RF cable to RF connector J1 and the other end to RF connector J4.
   e. Install power control cable-to-cable receptacle J2 and the other end to cable receptacle J5.

CAUTION: When making connection to the battery terminals of the vehicle, make sure connections are tight. Avoid accidental grounding of the positive terminal.

   f. Connect the DC power cable-to-cable receptacle J3 and the other end to the battery of the vehicle. If using power supply PP-7333, connect it to J1 of the power supply.
   g. Connect RF cable to RF connector J5 and the other end to RF connector J7.

WARNING: There is 1600 volts present at antenna terminal J4 when using the WHIP antenna or at the J3 terminal when using the 50-OHM antenna. Do not remove the cables during operation. Extreme caution must be taken to ensure that these terminals are at least 6 inches away from nearby objects, such as cables, guy wires, brackets, or ground leads.

   h. Attach the whip antenna lead-in cable to whip antenna connector J4 and the other end to the whip antenna-mounting base AB-652.
   i. When using doublet antenna, connect the 50-ohm coaxial cable to the 50-ohm connector J3 and the other end to the doublet antenna.
   j. Attach the ground strap to one of the terminals and the other end to the mount.
   k. Install the ground rod. Attach one end of the ground strap to one end of the terminal on front of antenna coupler and connect the other end to the ground rod.

3 - 56   16 February 2005

Performance Steps
    l. Connect the handset to top audio receptacle J1.

4. Install whip antenna.
    a. Assemble the whip antenna.
    b. Slide antenna cover over the assembled mast sections, tape the antenna tip assembly, and screw the antenna into the antenna mast base.
    c. Ensure antenna base is grounded securely.
    d. Tie the antenna down.
    e. Point the antenna in the direction of the distant station.

Performance Measures	GO	NO GO
1. Performed installation of the AM-6879 and the RT-1209.	——	——
2. Performed installation of the AM-6545 and the CU-2064.	——	——
3. Performed cable connections.	——	——
4. Installed whip antenna.	——	——

Evaluation Guidance: Score the Soldier a GO if all PMs are passed. Score the Soldier a NO-GO if any PM is failed. If the Soldier fails any PM, show what was done wrong and how to do it correctly. Have the Soldier perform the PMs until they are done correctly.

References
    Required   **Related**
    TM 11-5820-924-13

STP 11-25C13-SM-TG

Operate Radio Set AN/GRC-106(*)
113-620-2001

Conditions: Given a tactical or nontactical situation, a requirement, installed operational AN/GRC-106(*), TM 11-5820-520-10, and unit SOI.

NOTE: Supervision and assistance are available.

Standards: Within 20 minutes, conformed the meter readings on the radio set to the requirements in PMs 3 and 4, and placed the radio set into and out of operation according to PMs 1 through 6.

Performance Steps

DANGER: Dangerous voltages exist at the AM-3349/GRC-106 50-OHM LINE and WHIP antenna connectors. Be careful when working around the antenna or antenna connectors. Radio frequency voltages as high as 10,000 volts exist at these points. Operator and maintenance personnel should be familiar with the requirements of TB 43-0129 before attempting installation or operation of radio set AN/GRC-106(*). Injury or DEATH could result from improper or careless operation.

(Refer to TM 11-5820-520-10 for all performance steps, except as noted.)

1. Determine operating frequency. (Refer to unit SOI.)
2. Implement preliminary starting procedures.
3. Conduct starting procedures.
4. Conduct tuning procedures for voice or continuous wave operation.
5. Conduct voice or continuous wave operating procedures.
6. Perform stopping procedures.

Evaluation Preparation: Setup: Use the correct operating frequency.

Brief Soldier: This task must be completed without performing an unsafe act and without causing damage to the equipment or injury to you. (Sequence is scored).

Performance Measures	GO	NO GO

DANGER: Dangerous voltages exist at the AM-3349/GRC-106 50-OHM LINE and WHIP antenna connectors. Be careful when working around the antenna or antenna connectors. Radio frequency voltages as high as 10,000 volts exist at these points. Operator and maintenance personnel should be familiar with the requirements of TB 43-0129 before attempting installation or operation of radio set AN/GRC-106(*). Injury or DEATH could result from improper or careless operation.

(Refer to TM 11-5820-520-10 for all PMs, except as noted.)

	GO	NO GO
1. Determined operating frequency. (Refer to unit SOI.)	——	——
2. Implemented preliminary starting procedures.	——	——
3. Conducted starting procedures.	——	——
4. Conducted tuning procedures for voice or continuous wave operation.	——	——

STP 11-25C13-SM-TG

Performance Measures	GO	NO GO
5. Conducted voice or continuous wave operating procedures.	——	——
6. Performed stopping procedures.	——	——

Evaluation Guidance: Score the Soldier a GO if all PMs are passed. Score the Soldier a NO-GO if any PM is failed. If the Soldier fails any PM, show what was done wrong and how to do it correctly. Have the Soldier perform the PMs until they are done correctly.

References
    Required   **Related**
    TM 11-5820-520-10
    UNIT SOI

STP 11-25C13-SM-TG

Perform Operator Troubleshooting Procedures on Radio Set AN/GRC-106(*)
113-620-2002

Conditions: Given a tactical or nontactical situation, a requirement, radio set AN/GRC-106(*), DA PAM 738-750, TM 11-5820-520-10, and DA Form 2404 or DA Form 5988-E.

Standards: Corrected any discovered faults and recorded uncorrectable faults on DA Form 2404 or DA Form 5988-E, without error, and reported to your immediate supervisor according to PMs 1 through 3.

Performance Steps

DANGER: Dangerous voltages exist at the AM-3349/GRC-106 50-OHM LINE and WHIP antenna connectors. Be careful when working around the antenna or antenna connectors. Radio frequency voltages as high as 10,000 volts exist at these points. Operator and maintenance personnel should be familiar with the requirements of TB 43-0129 before attempting installation or operation of radio set AN/GRC-106(*). Injury or DEATH could result from improper or careless operation.

1. Conduct troubleshooting procedures on any abnormal condition while operating the AN/GRC-106(*) during normal operation or during operational check in daily PMCS. (Refer to TM 11-5820-520-10.)

2. List all uncorrectable faults, using the troubleshooting chart, on DA Form 2404 or DA Form 5988-E. (Refer to DA PAM 738-750.)

3. Notify immediate supervisor or supporting maintenance activity of any uncorrectable faults found. (Refer to DA PAM 738-750.)

Performance Measures	GO	NO GO
1. Conducted troubleshooting procedures on any abnormal condition while operating the AN/GRC-106(*) during normal operation or during operational check in daily PMCS. (Refer to TM 11-5820-520-10.)	——	——
2. Listed all uncorrectable faults, using the troubleshooting chart, on DA Form 2404 or DA Form 5988-E. (Refer to DA PAM 738-750.)	——	——
3. Notified immediate supervisor or supporting maintenance activity of any uncorrectable faults found. (Refer to DA PAM 738-750.)	——	——

Evaluation Guidance: Score the Soldier a GO if all PMs are passed. Score the Soldier a NO-GO if any PM is failed. If the Soldier fails any PM, show what was done wrong and how to do it correctly. Have the Soldier perform the PMs until they are done correctly.

References
Required
DA FORM 2404
DA FORM 5988-E
DA PAM 738-750
TM 11-5820-520-10

Related

Operate Radio Set AN/GRC-193A or Similar Radio Sets
113-620-2028

Conditions: Given an operational radio set AN/GRC-193A, operating frequency, a distant station, and TM 11-5820-924-13.

Standards: Performed all the proper procedures to place the radio set into operation and shut it down IAW TM 11-5820-924-13.

Performance Steps

1. Perform preliminary setup procedures.
    a. Ensure the air intake and exhaust ports on antenna coupler CU-2064 and power amplifier AM-6545 are clear.
    b. Ensure all cables are tight and connected to the proper connections on all components and equipment is grounded.
    c. Set the radio set controls.
        (1) OFF/MAX/VOLUME switch to OFF.
        (2) MODE to V-TR.
        (3) SB SELECTOR to USB or LSB.
        (4) 20 or 60MA (not used with VOICE or CW).
        (5) TTY/SPKR ON/OFF to OFF.
        (6) SQUELCH CONTROL to MID RANGE.
        (7) ANT CPLR CB - PUSH IN to RESET.
        (8) PA CB - PUSH IN to RESET.
        (9) 400W/100W selector to 100W.

2. Perform test procedures.
    a. Momentarily press the handset PUSH-TO-TALK (PTT) switch and verify that the handset emits a 1-kHz sidetone, which terminates in less than 12 seconds.
    b. Turn the radio off.

3. Set controls for operation.
    a. CIRCUIT BREAKER to ON.
    b. Start vehicle engine by adjusting throttle.
    c. OFF/MAX/VOLUME switch turn to right ½ turn.
    d. FREQ SELECT PUSH BUTTONS to set operating frequency.

4. Perform operating procedures. Wait for continuous 1-kHz tone to cease and receiver noise level to increase.

5. Perform shutdown procedures for radio set AN/GRC-193A.
    a. OFF/MAX/VOLUME switch to OFF.
    b. FREQ SELECTOR PUSH BUTTONS to ZEROIZE.
    c. PWR ON/OFF switch to OFF.
    d. CIRCUIT BREAKER to OFF.
    e. Vehicle THROTTLE push to IN.
    f. IGNITION switch to OFF.
    g. Disconnect whip antenna.

Performance Measures	**GO**	**NO GO**
1. Performed preliminary setup procedures.	——	——
2. Performed test procedures.	——	——
3. Set controls for operation.	——	——

STP 11-25C13-SM-TG

Performance Measures <u>GO</u> <u>NO GO</u>

4. Performed operating procedures. Waited for continuous 1-kHz tone to cease and receiver noise level to increase. —— ——

5. Performed shutdown procedures for radio set AN/GRC-193A. —— ——

Evaluation Guidance: Score the Soldier a GO if all PMs are passed. Score the Soldier a NO-GO if any PM is failed. If the Soldier fails any PM, show what was done wrong and how to do it correctly. Have the Soldier perform the PMs until they are done correctly.

References
   Required   **Related**
   TM 11-5820-924-13

STP 11-25C13-SM-TG

Operate Radio Set AN/PRC-150 Radio Set in Single-Channel Mode
113-620-2051

Conditions: Given Harris RT 150 and Publication Number 10515-0103-4100.

Standards: Entered the radio set into the net and the Soldier made a radio check.

Performance Steps

1. Prepare radio set for operation.
    a. Install two batteries BA-5590/U or BB-590/U into Battery Box.
    b. Connect Battery Box to the radio set.
    c. Connect the handset to the RT's front panel J1 audio connector.
    d. Connect the OE-505 whip antenna to the J7 front panel antenna connector.

2. Perform power-up procedures.
    a. Rotate the function switch to the PT position.
    b. R/T will run a self-test.
    c. The RT automatically begins to scan, listening for its unique self-address to be called.

3. Perform initial checks. (Refer to para 3.3.1 in Publication Number 10515-0103-4100.)

4. Load keys into the radio set.

NOTE: The following steps for loading the radio set are used with the AN/CYZ-10. If using another load device, refer to para 3.12 in Publication Number 10515-0103-4100.

   a. Turn AN/CYZ-10 DTD fill device on.
   b. Connect fill device to J18 FILL connector when directed by the DTD menu instructions.
   c. Select the key to be loaded and use ISSUE as the transmit mode.
   d. Rotate function switch to LD.
   e. Select KYK-13 as the fill device and press [ENT].
   f. Select the KEY TYPE, then select the key compartment position number (01-25). Press [ENT].

NOTE: KEKs and TrKEKs do not require a compartment number.

   g. INITIATE FILL AT FILL DEVICE displays.
   h. Press SEND on the DTD. The radio display FILL IN PROGRESS.
   i. When FILL DONE displays press [ENT].
   j. At prompt MORE FILL DATA? select YES to enter more fill data. Repeat steps e through i. Use the DTD menu to back up for new key selection.
   k. When all fill data is entered, select NO when the MORE FILL DATA? prompt displays.
   l. Turn off DTD and disconnect it from the J18 FILL connector.
   m. Rotate function switch from LD to PT, CT, or CC.

5. Perform operational procedures to operate in the single channel mode.
    a. Rotate the function switch to the PT position.
    b. Press the MODE.
    c. Select the FIX operating mode.
    d. Press ENT.
    e. Select desired frequency by pressing PRE +/-.
    f. Press PTT on the handset to initiate automatic tuning.
    g. Rotate the function switch to the PT, CT, or CC position.
    h. Make a radio check.
    i. Continue with operations.

STP 11-25C13-SM-TG

Evaluation Preparation: Setup: For this evaluation, prepare equipment and ensure equipment is complete and operational.

Brief Soldier: Tell the Soldier that all PMs are performed in sequence and must be passed.

Performance Measures	GO	NO GO
1. Prepared radio set for normal operations.	——	——
2. Performed power-up procedures.	——	——
3. Performed initial checks.	——	——
4. Loaded keys into radio set.	——	——
5. Preformed operational procedures to operate in the single channel mode.	——	——
6. Conducted a radio check.	——	——

Evaluation Guidance: Score the Soldier a GO if all PMs are passed. Score the Soldier a NO-GO if any PM is failed. If the Soldier fails any PM, show what was done wrong and how to do it correctly. Have the Soldier perform the PMs until they are done correctly.

References
   Required   **Related**
   PUB # 10515-0103-4100

STP 11-25C13-SM-TG

Operate AN/PRC-150() in Automatic Link Establishment(ALE) Mode
113-620-2052

Conditions: Given a preprogrammed radio set and Publication Number 10515-0103-4100.

Standards: The Soldier placed an ALE call.

Performance Steps

1. Prepare radio set for operation.
    a. Install two batteries BA-5590/U or BB-590/U into battery box.
    b. Connect battery box to radio set. Refer to figure 2-4 in Publication Number 10515-0103-4100.
    c. Connect handset to radio transmitter (R/T) front panel J1 AUDIO connector.
    d. Connect the OE-505 whip antenna to the J7 FRONT ANTENNA connector on the front panel of the R/T.

2. Perform power-up procedures.
    a. Rotate the function switch to the PT position.
    b. R/T will run self-test.
    c. The RT automatically begins to scan, listening for its unique self-address to be called.

3. Perform initial checks. (Refer to para 3.3.1 in Publication Number 10515-0103-4100.)

4. Load keys into the radio set.

NOTE: The following steps for loading the radio set are used with the AN/CYZ-10. If using another load device, refer to para 3.12 in Publication Number 10515-0103-4100.

    a. Turn AN/CYZ-10 DTD fill device on.
    b. Connect fill device to J18 FILL connector when directed by the DTD menu instructions.
    c. Select the key to be loaded and use ISSUE as the transmit mode.
    d. Rotate function switch to LD.
    e. Select KYK-13 as the fill device and press [ENT].
    f. Select the KEY TYPE, then select the key compartment position number (01-25). Press [ENT].

NOTE: KEKs and TrKEKs do not require a compartment number.

    g. INITIATE FILL AT FILL DEVICE displays.
    h. Press SEND on the DTD. The radio display FILL IN PROGRESS.
    i. When FILL DONE displays press [ENT].
    j. At prompt MORE FILL DATA? select YES to enter more fill data. Repeat steps e through i. Use the DTD menu to back up for new key selection.
    k. When all fill data is entered, select NO when the MORE FILL DATA? prompt displays.
    l. Turn off DTD and disconnect it from the J18 FILL connector.
    m. Rotate function switch from LD to CC.

5. Perform operational procedures to operate in the ALE mode.
    a. Press [MODE].
    b. Select ALE.

NOTE: The radio cannot be placed in ALE mode until ALE programming has taken place.

    c. The R/T automatically beings to scan, listening for its unique self-address to be called.

6. Place an ALE call.
    a. Press the call [CALL] button.
    b. Press the arrow buttons to select AUTOMATIC and press [ENT].
    c. Press the arrows keys to select the address type and press [ENT].

NOTE: For this task, select ANY as address type.

STP 11-25C13-SM-TG

Performance Steps
    d. The R/T will transmit to the selected address. After the call, the R/T will wait for the response.
    e. The R/T will indicate LINKED once a response is received.
    f. Press [CLR] to return to the system preset screen.

Performance Measures	**GO**	**NO GO**
1. Installed batteries.	——	——
2. Performed power-up procedures.	——	——
3. Performed initial checks.	——	——
4. Loaded keys into the radio set.	——	——
5. Performed operational procedures to operate in the ALE mode.	——	——
6. Placed an ALE call.	——	——

Evaluation Guidance: Score the Soldier a GO if all PMs are passed. Score the Soldier a NO-GO if any PM is failed. If the Soldier fails any PM, show what was done wrong and how to do it correctly. Have the Soldier perform the PMs until they are done correctly.

References
    Required   **Related**
    PUB # 10515-0103-4100

STP 11-25C13-SM-TG

Perform PMCS on Radio Set AN/GRC-106(*)
113-620-3001

Conditions: Given a tactical or nontactical situation; a requirement; radio set AN/GRC-106(*); mild detergent solution; cleaning fluid trichloroethane; clean, dry lint-free cloth; DA PAM 738-750, TM 11-5820-520-10, and DA Form 2404 or DA Form 5988-E.

NOTE: Supervision and assistance are available.

Standards: Cleaned the exteriors of all components, tightened all dials and knobs, all cables were in good condition (no cracks or broken connectors), all fuses were of correct value, and recorded uncorrectable faults on DA Form 2404 or DA Form 5988-E, without error, and reported to your immediate supervisor according to PMs 1 through 3.

Performance Steps

DANGER: Dangerous voltages exist at the AM-3349/GRC-106 50-OHM LINE and WHIP antenna connectors. Be careful when working around the antenna or antenna connectors. Radio frequency voltages as high as 10,000 volts exist at these points. Operator and maintenance personnel should be familiar with the requirements of TB 43-0129 before attempting installation or operation of radio set AN/GRC-106(*). Injury or DEATH could result from improper or careless operation.

1. Perform PMCS at the required intervals. (Refer to TM 11-5820-520-10.)

2. Complete DA Form 2404 or DA Form 5988-E as a daily maintenance form. (Refer to DA PAM 738-750.)

3. Report all uncorrectable defects. (Refer to DA PAM 738-750.)
    a. Notify immediate supervisor of all uncorrectable faults found.
    b. Submit DA Form 2404 or DA Form 5988-E to supervisor or support maintenance personnel.

Evaluation Preparation: Setup: Clean the exterior surfaces of the units comprising the AN/GRC-106(*).

Brief Soldier: Tell the Soldier all steps must be performed correctly to receive a GO. (Sequence is not scored.)

Performance Measures	GO	NO GO
DANGER: Dangerous voltages exist at the AM-3349/GRC-106 50-OHM LINE and WHIP antenna connectors. Be careful when working around the antenna or antenna connectors. Radio frequency voltages as high as 10,000 volts exist at these points. Operator and maintenance personnel should be familiar with the requirements of TB 43-0129 before attempting installation or operation of radio set AN/GRC-106(*). Injury or DEATH could result from improper or careless operation.		
1. Performed PMCS at the required intervals. (Refer to TM 11-5820-520-10.)	——	——
2. Completed DA Form 2404 or DA Form 5988-E as a daily maintenance form. (Refer to DA PAM 738-750.)	——	——

3. Reported all uncorrectable defects. (Refer to DA PAM 738-750.)
  a. Notified immediate supervisor of all uncorrectable faults found.
  b. Submitted DA Form 2404 or DA Form 5988-E to supervisor or support maintenance personnel.
	——	——

Evaluation Guidance: Score the Soldier a GO if all PMs are passed. Score the Soldier a NO-GO if any PM is failed. If the Soldier fails any PM, show what was done wrong and how to do it correctly. Have the Soldier perform the PMs until they are done correctly.

References
  Required   **Related**
  DA FORM 2404
  DA FORM 5988-E
  DA PAM 738-750
  TM 11-5820-520-10

STP 11-25C13-SM-TG

Perform Unit Level Maintenance for AN/GRC-193A or Similar Radio Sets
113-620-3052

Conditions: Given a tactical or nontactical situation, under all weather conditions; a requirement; radio set AN/GRC-193A, complete; TM 11-5820-924-13, DA PAM 738-750, DA Form 2404 or DA Form 5988-E; soft, damp cloth and a soft-bristled brush. This task may be performed in an NBC environment.

Standards: Performed PMCS, correctly identified two of three faults, and completed DA Form 2404 or DA Form 5988-E IAW TM 11-5820-924-13.

Performance Steps

DANGER: Dangerous voltages exist at the CU-2064, 50-OHM LINE, and WHIP antenna connectors. Be careful when working around the antenna connectors. Radio frequency voltages as high as 10,000 volts exist at these points. Operator and maintenance personnel should be familiar with the requirements of TB SIG 43-0129 before attempting installation or operation of radio set AN/GRC-193A. Injury or DEATH could result from failure to comply with safe practices.

1. Perform daily checks and services.
    a. Check cover-fastening screws on the power amplifier and the antenna coupler for tightness.
    b. Ensure all units are securely clamped to the mount.
    c. Ensure all units and cables are undamaged and all cable connections are tight.
    d. Ensure rubber seals on the power amplifier circuit breakers are not damaged.
    e. Clean the external surfaces of all units.
    f. Remove anything blocking the air inlets (front) and air exhausts (rear) of the power amplifier and antenna coupler.

2. Perform weekly checks and services or before each time the radio set is used.

CAUTION: Do not clean exterior surfaces if the power is ON.

   a. Clean exterior surfaces of the units of the AN/GRC-193A.
   b. Check all interconnecting cables and connectors for cracks and breaks. Ensure the bonding jumper grounds the equipment. Replace cables that have cuts, cracks, or broken connectors.
   c. Ensure the meter faces (glass) are not loose or broken.
   d. Check fuses on the AM-6879 for correct value. Check spare for quantity and proper value.
   e. Ensure there are no items in front or back of the CU-2064 and AM-6545 that will obstruct airflow through the intake louver.
   f. Ensure rubber seals on the AM-6545 circuit breakers are not damaged or missing.

3. Perform weekly checks and services or during each time the radio set is used.
    a. During operational checks, observe that the mechanical action of knobs, switches, and controls is smooth and free of external or internal binding.
    b. Operate the equipment on an authorized frequency to verify its capabilities.

4. Perform troubleshooting procedures.
    a. Use audible tones to isolate a fault.
    b. Use performance test to isolate a fault.
    c. Use flowchart and troubleshooting guide to isolate a fault.

5. Complete DA Form 2404 or DA Form 5988-E.

Performance Measures	**GO**	**NO GO**
1. Performed daily checks and services.	——	——
2. Performed weekly checks and services or before each time the radio set was used.	——	——

STP 11-25C13-SM-TG

Performance Measures <u>GO</u> <u>NO GO</u>

3. Performed weekly checks and services or during each time the radio set was used. ____ ____

4. Performed troubleshooting procedures. ____ ____

5. Completed DA Form 2404 or DA Form 5988-E. ____ ____

Evaluation Guidance: Score the Soldier a GO if all PMs are passed. Score the Soldier a NO-GO if any PM is failed. If the Soldier fails any PM, show what was done wrong and how to do it correctly. Have the Soldier perform the PMs until they are done correctly.

References
    Required   **Related**
    DA FORM 2404
    DA FORM 5988-E
    DA PAM 738-750
    TM 11-5820-924-13

STP 11-25C13-SM-TG

Subject Area 6: COMBAT NET RADIOS

Prepare SINCGARS (Manpack) for Operation
113-587-1064

Conditions: Given an operational SINCGARS manpack radio with battery box CY-8323A/B, battery, antenna AS-3683, handset H-250, carrying case, and TM 11-5820-890-10-8.

Standards: Correctly installed the battery, preset the function controls for operation, mounted the radio in the backpack, and correctly connected accessories.

Performance Steps

1. Install Battery.
    a. Install one battery used for main power in the SIP radio (Rechargeable, BB-390 A/U Battery; Rechargeable, BB-590/U; or Non-Rechargeable, BB-5590/U (Lithium).
    b. Connect battery case to radio set. (Refer to TM 11-5820-890-10-8, page 2-14.)

2. Assemble radio and pack frame.
    a. Assemble pack and fasten assembled radio into carrying case. (Refer to TM 11-5820-890-10-8.)

3. Install antenna.
    a. Connect AS-3683/PRC to radio. (Refer to TM 11-5820-890-10-8, page 5-3.)

4. Connect handset H-250.
    a. Connect handset H-250/U or Handheld Remote Control Radio Device (HRCRD) (C-12493/U). (Refer to TM 11-5820-890-10-8, page 3-12.)

5. Preset function controls. (Refer to TM 11-5820-890-10-8, page 5-1.)

Performance Measures	GO	NO GO
1. Installed batteries.	——	——
2. Assembled radio and pack frame.	——	——
3. Installed antenna.	——	——
4. Connected handset H-250.	——	——
5. Preset function controls.	——	——

Evaluation Guidance: Score the Soldier a GO if all PMs are passed. Score the Soldier a NO-GO if any PM is failed. If the Soldier fails any PM, show what was done wrong and how to do it correctly. Have the Soldier perform the PMs until they are done correctly.

References
Required
TM 11-5820-890-10-8

Related
TM 11-5820-890-10-1
TM 11-5820-890-10-3

STP 11-25C13-SM-TG

Operate SINCGARS Single-Channel (SC)
113-587-2070

Conditions: Given an operational SINCGARS, KYK-13/TSEC with keys or AN/CYZ-10, C-11291 CM, distant station, TM 11-5820-890-10-8, TM 11-5820-890-10-3, TM 11-5820-890-10-1, ACP 125 US Suppl-1, DA PAM 738-750, FM 24-19, FM 24-18, and unit SOI or ANCD with SOI data loaded.

Standards: Conducted a secure communications check in SC mode with a distant station and changed the radio functions using the control monitor.

Performance Steps

1. Perform starting procedures.
2. Load traffic encryption key (TEK).
3. Enter net.
    a. Use correct procedures.
    b. Conduct secure communications check.
4. Prepare control monitor for operation.
5. Change radio functions using the control monitor.
6. Exit net.
7. Perform stopping procedures.

Evaluation Preparation: Setup: Ensure radio set is complete and operational with CM installed on radio.

Brief Soldier: Tell the Soldier all PMs must be completed correctly and done in sequence within 20 minutes.

Performance Measures	GO	NO GO
1. Performed starting procedures.	——	——
2. Loaded TEK.	——	——
3. Entered net.	——	——
4. Prepared control monitor for operation.	——	——
5. Changed radio functions using the control monitor.	——	——
6. Exited the net.	——	——
7. Performed stopping procedures.	——	——

Evaluation Guidance: Score the Soldier a GO if all PMs are passed. Score the Soldier a NO-GO if any PM is failed. If the Soldier fails any PM, show what was done wrong and how to do it correctly. Have the Soldier perform the PMs until they are done correctly.

References

Required	**Related**

ACP 125 US SUPPL-1
DA PAM 738-750
FM 24-18
FM 24-19
TM 11-5820-890-10-1
TM 11-5820-890-10-3
TM 11-5820-890-10-8
UNIT SOI

STP 11-25C13-SM-TG

Operate SINCGARS Frequency Hopping (FH) (Net Members)
113-587-2071

Conditions: Given an operational SINCGARS radio, ECCM fill device with FH data, KYK-13/TSEC with keys, C-11291 CM, distant net control station (NCS), unit SOI, DA Form 2404, TM 11-5820-890-10-8, TM 11-5820-890-10-3, TM 11-5820-890-10-1, ACP 125 US Suppl-1, DA PAM 738-750, FM 24-19, and FM 24-18.

Standards: Established FH communications using the cold start and CUE late net entry methods, successfully completed the radio check, and changed the radio functions using the control monitor.

Performance Steps

1. Perform starting procedures. (Set radio to PLGR (AN/PSN-11) time.)

2. Perform net member cold start procedures.
    a. Use correct call signs.
    b. Use correct procedures.

3. Perform net member CUE late net entry.
    a. Use correct call signs.
    b. Use correct procedures.

4. Prepare control monitor for operation.

5. Change radio functions using the control monitor.

6. Perform stopping procedures.

Evaluation Preparation: Setup: Ensure radio set is complete and operational with CM installed on radio set.

Brief Soldier: Tell the Soldier all PMs must be completed correctly and done in sequence within 20 minutes.

Performance Measures	GO	NO GO
1. Performed starting procedures. (Set Radio to PLGR (PSN-11) time.)	——	——
2. Performed net member cold start procedures.	——	——
3. Performed net member CUE late net entry.	——	——
4. Prepared control monitor for operation.	——	——
5. Changed radio functions using the control monitor.	——	——
6. Performed stopping procedures.	——	——

Evaluation Guidance: Score the Soldier a GO if all PMs are passed. Score the Soldier a NO-GO if any PM is failed. If the Soldier fails any PM, show what was done wrong and how to do it correctly. Have the Soldier perform the PMs until they are done correctly.

References
    Required   **Related**
    ACP 125 US SUPPL-1
    DA FORM 2404
    DA PAM 738-750
    FM 24-18
    FM 24-19
    TM 11-5820-890-10-1
    TM 11-5820-890-10-3
    TM 11-5820-890-10-8
    UNIT SOI

STP 11-25C13-SM-TG

Operate SINCGARS Frequency Hopping (FH) Net Control Station (NCS)
113-587-2072

Conditions: Given an operational SINCGARS, ECCM fill device with FH data, KYK-13/TSEC with keys, distant stations, TM 11-5820-890-10-1, TM 11-5820-890-10-3, ACP 125 US Suppl-1, DA PAM 738-750, FM 24-19, FM 24-18, and unit SOI.

Standards: Sent an operational message and received in the FH mode using the cold start and CUE late net entry procedures.

Performance Steps

1. Perform starting procedures. (Set Radio to PLGR (AN/PSN-11) time.)

2. Perform NCS permission checks.

3. Perform NCS cold start net opening.
    a. Use correct call signs.
    b. Conduct FH communications.

4. Perform NCS CUE late net entry.
    a. Use correct call signs.
    b. Conduct FH communications.

5. Perform stopping procedures.

Performance Measures	GO	NO GO
1. Performed starting procedures. (Set radio to PLGR (AN/PSN-11) time.)	——	——
2. Performed NCS permission checks.	——	——
3. Performed NCS cold start net opening.	——	——
4. Performed NCS CUE late net entry.	——	——
5. Performed stopping procedures.	——	——

Evaluation Guidance: Score the Soldier a GO if all PMs are passed. Score the Soldier a NO-GO if any PM is failed. If the Soldier fails any PM, show what was done wrong and how to do it correctly. Have the Soldier perform the PMs until they are done correctly.

References

Required
ACP 125 US SUPPL-1
DA PAM 738-750
FM 24-18
FM 24-19
TM 11-5820-890-10-1
TM 11-5820-890-10-3
UNIT SOI

Related
TM 11-5820-890-10-8

STP 11-25C13-SM-TG

Operate SINCGARS Remote Control Unit (SRCU)
113-587-2077

Conditions: Given an SINCGARS ICOM operating in a net, SRCU C-11561, battery BA-1372, battery BA-5590, battery case CY-8346, flat-tip screwdriver, installed field wire line WF-16, pack frame, cable CX-13298, distant station, TM 11-5820-890-10-1, and TM 11-5820-890-20-1.

Standards: Sent and received an operational message from the SRCU.

Performance Steps

1. Prepare SRCU for operation.
2. Operate SRCU single channel (SC).
3. Operate SRCU frequency hopping (FH).

Performance Measures	GO	NO GO
1. Prepared SRCU for operation.	——	——
2. Operated SRCU SC.	——	——
3. Operated SRCU FH.	——	——

Evaluation Guidance: Score the Soldier a GO if all PMs are passed. Score the Soldier a NO-GO if any PM is failed. If the Soldier fails any PM, show what was done wrong and how to do it correctly. Have the Soldier perform the PMs until they are done correctly.

References

Required	Related
TM 11-5820-890-10-1	
TM 11-5820-890-20-1	

STP 11-25C13-SM-TG

Troubleshoot Secure Single-Channel Ground and Airborne Radio Systems (SINCGARS) ICOM With or Without the AN/VIC-1 or AN/VIC-3
113-587-0070

Conditions: Given an inoperable secure SINCGARS with or without intercommunications set AN/VIC-1, with or without frequency-hopping multiplexer (FHMUX) TD-1456/VRC, operating in a net; digital multimeter AN/PSM-45(*); radio test set AN/PRM-34; tool kit TK-101/G; ANCD AN/CYZ-10 with fill, fill cable; TM 11-5820-890-20-1, TM 11-5820-890-20-2, or TM 11-5820-890-20-3; DA Form 5986-E (Preventive Maintenance Schedule and Record [EGA]), and DA Form 5988-E.

Standards: Restored the radio set to operation or evacuated the defective LRU to a higher maintenance level.

Performance Steps

1. Verify reported malfunctions.
    a. Review operator's actions.
    b. Perform visual inspections.

2. Perform systematic troubleshooting procedures.
    a. Perform self-test.
    b. Perform off-line measurements.

3. Take corrective actions.
    a. Fix or replace the defective LRU. (Refer to the MAC.)
    b. Perform an operational check.
    c. Evacuate the defective LRU to a higher maintenance level.
        (1) Process for a higher maintenance level.
        (2) Prepare maintenance forms.

Performance Measures   <u>GO</u>   <u>NO GO</u>
(Refer to TM 11-5820-890-20-1, TM 11-5820-890-20-2, or TM 11-5820-890-20-3 for all PMs.)

1. Verified reported malfunctions.   ____   ____
    a. Reviewed operator's actions.
    b. Performed visual inspection.

2. Performed systematic troubleshooting procedures.   ____   ____
    a. Performed self-test.
    b. Performed off-line measurements.

3. Took corrective actions.   ____   ____
    a. Fixed or replaced the defective LRU. (Refer to the MAC.)
    b. Performed an operational check.
    c. Evacuated the defective LRU to a higher maintenance level.
        (1) Processed for a higher maintenance level.
        (2) Prepared maintenance forms.

Evaluation Guidance: Score the Soldier a GO if all PMs are passed. Score the Soldier a NO-GO if any PM is failed. If the Soldier fails any PM, show what was done wrong and how to do it correctly. Have the Soldier perform the PMs until they are done correctly.

References
   Required
   DA FORM 5986-E
   DA FORM 5988-E
   TM 11-5820-890-20-1
   TM 11-5820-890-20-2
   TM 11-5820-890-20-3

   Related
   (O)TM 11-5810-394-14&P
   DA PAM 738-750
   FM 11-32
   TM 11-6625-3015-14
   TM 11-6625-3199-14

STP 11-25C13-SM-TG

Troubleshoot Single-Channel Demand Assigned Multiple Access (DAMA) Tactical Satellite Terminal AN/PSC-5
113-589-0045

Conditions: Given radio set AN/PSC-5, digital multimeter AN/PSC-45, the ANCD with fill, fill cable, tool kit TK-101/G, TM 11-5820-1130-12&P, (O)TM 11-5810-394-14&P, and the technical manual for the ANCD.

Standards: The AN/PSC-5 radio set processed operational traffic; or evacuated the defective LRU to a higher maintenance level.

Performance Steps
(Refer to TM 11-5820-1130-12&P for all performance steps.)

1. Verify reported malfunctions.
    a. Review operator's actions.
    b. Perform visual inspection.

2. Perform systematic troubleshooting procedures.
    a. Sectionalize the systems. (Determine inoperable section.)
    b. Localize inoperable station. (Perform diagnostics.)
    c. Isolate the defective LRU.

3. Take corrective actions.
    a. Fix or replace the defective LRU. (Refer to MAC.)
    b. Perform an operational check.
    c. Evacuate the LRU with uncorrectable deficiencies to a higher maintenance level.
        (1) Process for a higher maintenance level.
        (2) Prepare maintenance records.

Performance Measures	**GO**	**NO GO**
(Refer to TM 11-5820-1130-12&P for all PMs.)		

1. Verified reported malfunctions.
    a. Reviewed operator's actions.
    b. Performed visual inspection.
	——	——

2. Performed systematic troubleshooting procedures.
    a. Sectionalized the systems. (Determined inoperable section.)
    b. Localized inoperable station. (Performed diagnostics.)
    c. Isolated the defective LRU.
	——	——

3. Took corrective actions.
    a. Fixed or replaced the defective LRU. (Refer to the MAC.)
    b. Performed an operational check.
    c. Evacuated the LRU with uncorrectable deficiencies to a higher maintenance level.
        (1) Processed for a higher maintenance level.
        (2) Prepared maintenance records.
	——	——

Evaluation Guidance: Score the Soldier a GO if all PMs are passed. Score the Soldier a NO-GO if any PM is failed. If the Soldier fails any PM, show what was done wrong and how to do it correctly. Have the Soldier perform the PMs until they are done correctly.

References
   Required   **Related**
   (O)TM 11-5810-394-14&P
   ANCD TM
   TM 11-5820-1130-12&P

STP 11-25C13-SM-TG

Operate Secure Single-Channel Tactical Satellite Radio Set in SATCOM Mode
113-589-2028

Conditions: As a Radio Operator-Maintainer in a field environment, given AN/CYZ-10(*) Automated Net Control Device (ANCD), TACSAT terminal AN/PSC-5 or similar TACSAT Radio Set, satellite antenna AS-4326/P, battery box, two BB-590/U batteries, fill cable, and TM 11-5820-1130-12&P.

Standards: The Soldier established communications using SATCOM mode of operation.

Performance Steps

1. Prepare radio set for operation.
    a. Install two Batteries (BA-5590/U or BB-390A) into battery case.
    b. Connect battery case to radio set.
    c. Connect Handset H-250()/U to AUDIO connector.
    d. Install antenna.
        (1) Line-of-Sight (LOS) Antenna.

WARNING: Physical contact with any nearby metallic objects may cause a RF shock or burn. Do not use the antenna if the sheath covering is damaged or removed because contact with the internal metallic parts of the antenna can cause a RF shock or burn.

   (a) Loosen friction ring on LOS antenna.
   (b) Connect LOS antenna to ANTENNA connector of R/T by turning fully clockwise.
   (c) Position LOS antenna as desired and tighten friction ring.
        (2) Satellite antenna.

WARNING: Satellite communications antennas concentrate transmitter signals into beams of high-energy electromagnetic radiation. Do not stand in front of the satellite antenna or touch it any time when transmitting.

   (a) Remove antenna from carrying case.
   (b) Release leg strap holding tripod legs around antennas assembly. Slide leg strap off leg and stow in carrying case.
   (c) Pull out and swing tripod legs to receptacles and set up on ground.
   (d) Press down on locking ring and release found dipole elements.
   (e) Open eight telescoping ground plan arms by pulling down and out on the conductive chain attached to each arm.
   (f) Loosen "T" screw. Adjust antenna for desired elevation angle and hand tighten "T" screw.
   (g) Loosen "T" screw. Position dipole elements over a tripod leg. Hand tighten "T" screw.
   (h) Using your compass, position antenna to desired azimuth according to operational requirements.
   (i) Unfold radials; then, connect two radials together as an array.
   (j) Insert long end of array into the driven assembly. Ensure the array elements align with dipole elements.
   (k) Connect P2 of antenna cable W6 to satellite antenna connector. Connect P1 of W6 to ANT connector on R/T.

2. Power-up procedure
    a. Set MODE switch to PT or CT position. The radio set displays "AN/PSC-5 Initializing Radio" for approximately 2 seconds. Then the message "AN/PSC-5 Initializing Modules" will appear for 1 second or less.

NOTE: Selecting CT will not run BIT.

   b. Observe that radio set initiates power-up BIT. Power-up BIT will be completed after approximately 30 seconds. The display will show the last active CURRENT MODE menu.

STP 11-25C13-SM-TG

Performance Steps

NOTE: If power-up BIT fails, refer to TM 11-5820-1130-12&P, Table 5-3.

    3. Load COMSEC. (Refer to TM 11-5820-1130-12&P, para 4.13.)

    4. Load SATCOM Presets. (Refer to TM 11-5820-1130-12&P, para 4.23.2 for procedures.)

    5. SATCOM Operation.
        a. Set SQUELCH and VOLUME controls to mid-range.
        b. Verify SATCOM is selected on CURRENT MODE menu. If not selected, press NEXT/PREV keys to move cursor to mode field. Press arrow keys until SATCOM is displayed. Then press ENT key.

NOTE: Whenever the ENT key is pressed while in the CURRENT MODE menu, the menu title will alternate with the message CONFIGURING: WAIT while the parameters are being processed. Wait until the CONFIGURING WAIT message is no longer displayed before proceeding with operation.

        c. Select desired operating preset. Press ENT key.
        d. If in CT, proceed to next step. If in PT, proceed to step (4).
            (1) While observing filed strength indication on display (Sq- 040), position satellite antenna for maximum field strength indication.
            (2) To transmit, press and hold PTT switch on handset while talking in the mouthpiece (wait for single beep in handset before transmitting). Observe that transmit indication (Tx-CT-130) and signal strength (Tx-CT-130) is shown on display during transmission.
            (3) To receive, release PTT switch and listen to the handset earpiece. Observe that receive indication (Rx-CT-170) and signal strength (Rx-CT-170) is shown on display during reception.
            (4) Connect a data device. (If using KL-43C/KL-43F, refer to TM 11-5820-1130-12&P, para 4.8.1; if using Digital Communications Terminal (DCT) AN/PSC-2, refer to para 4.8.2; if using Digital Message Device Group (DMDG) OA-8990, refer to para 4.8.3.)
                (a) Send message by keying data device. Observe transmit indication on radio set.
                (b) Receive message on data device. Observe receive indication on radio set.

Evaluation Preparation: Setup: Ensure that all information, references, and equipment required to perform the task are available. Use the TMs and the evaluation guide to score the Soldier's performance.

Brief the Soldier. Tell the Soldier what he is required to do IAW the task conditions and standards.

Performance Measures	**GO**	**NO GO**
1. Prepared radio set for operation.	——	——
2. Performed power-up procedures.	——	——
3. Loaded keys into the radio set.	——	——
4. Loaded SATCOM presets.	——	——
5. Sent/received communications using SATCOM mode of operation	——	——

Evaluation Guidance: Score the Soldier a GO if all PMs are passed. Score the Soldier a NO-GO if any PM is failed. If the Soldier fails any PM, show what was done wrong and how to do it correctly. Have the Soldier perform the PMs until they are done correctly.

References
    Required　　　　　　　　　　　　　　　**Related**
    TM 11-5820-1130-12&P

STP 11-25C13-SM-TG

Operate Secure Single-Channel Tactical Satellite Radio Set in Demand Assigned Multiple Access (DAMA) Mode
113-589-2029

Conditions: As a Radio Operator-Maintainer in a field environment, given AN/CYZ-10(*) Automated Net Control Device (ANCD), TACSAT terminal AN/PSC-5 or similar TACSAT radio set, satellite antenna AS-4326/P, battery box, two BB-5990/U batteries, fill cable, and TM 11-5820-1130-12&P.

Standards: The Soldier established communications using SATCOM mode of operation.

Performance Steps

1. Prepare radio set for operation.
    a. Install two Batteries (BA-5590/U or BB-390A) into battery case.
    b. Connect battery case to radio set.
    c. Connect Handset H-250()/U to AUDIO connector.
    d. Install antenna.
        (1) Line-of-Sight (LOS) Antenna.

WARNING: Physical contact with any nearby metallic objects may cause a RF shock or burn. Do not use the antenna if the sheath covering is damaged or removed because contact with the internal metallic parts of the antenna can cause a RF shock or burn.

   (a) Loosen friction ring on LOS antenna.
   (b) Connect LOS antenna to ANTENNA connector of R/T by turning fully clockwise.
   (c) Position LOS antenna as desired and tighten friction ring.
        (2) Satellite antenna.

CAUTION: Satellite communications antennas concentrate transmitter signals into beams of high-energy electromagnetic radiation. Do not stand in front of the satellite antenna or touch it any time when transmitting.

   (a) Remove antenna from carrying case.
   (b) Release leg strap holding tripod legs around antennas assembly. Slide leg strap off leg and stow in carrying case.
   (c) Pull out and swing tripod legs to receptacles and set up on ground.
   (d) Press down on locking ring and release found dipole elements.
   (e) Open eight telescoping ground plan arms by pulling down and out on the conductive chain attached to each arm.
   (f) Loosen "T" screw. Adjust antenna for desired elevation angle and hand tighten "T" screw.
   (g) Loosen "T" screw. Position dipole elements over a tripod leg. Hand tighten "T" screw.
   (h) Using your compass, position antenna to desired azimuth according to operational requirements.
   (i) Unfold radials; then, connect two radials together as an array.
   (j) Insert long end of array into the driven assembly. Ensure the array elements align with dipole elements.
   (k) Connect P2 of antenna cable W6 to satellite antenna connector. Connect P1 of W6 to ANT connector on R/T.

2. Power-up procedures.
    a. Set MODE switch to CT position. The radio set displays "AN/PSC-5 Initializing Radio" for approximately 2 seconds. Then the message "AN/PSC-5 Initializing Modules" will appear for 1 second or less.

NOTE: Selecting CT will not run BIT.

STP 11-25C13-SM-TG

Performance Steps

    b. Observe that radio set initiates power-up BIT. Power-up BIT will be completed after approximately 30 seconds. The display will show the last active CURRENT MODE menu.

NOTE: If power-up BIT fails, refer to TM 11-5820-1130-12&P, Table 5-3.

3. Load COMSEC. (Refer to TM 11-5820-1130-12&P, para 4.13.)

4. Load Orderwire Keys. (Refer to TM 11-5820-1130-12&P, para 4.14.)

NOTE: If using a radio that is cloned, performance steps 5 thru 8 are not required.

5. Configure terminal.
    a. Load Guard List. (Refer to TM 11-5820-1130-12&P, para 4.26.2.)
    b. Load Satellite Ephemeris. (Refer to TM 11-5820-1130-12&P, para 4.26.5.)
    c. Load Information Codes. (Refer to TM 11-5820-1130-12&P, para 4.26.6.)
    d. Load Terminal data. (Refer to TM 11-5820-1130-12&P, para 4.26.3.)
    e. Load I/O rates and setup. (Refer to TM 11-5820-1130-12&P, para 4.26.4.)

6. Load Pre-mission Demand Assignment Multiple Access (DAMA) setup for 5 kHz service.
    a. Load I/O rates and setup. (Refer to TM 11-5820-1130-12&P, para 4.26.4.)
    b. From the MAIN MENU, press the 3 key. The SET PRESETS menu will display.
    c. Press the 2 key. The SET PRESET 5K menu will display.
    d. With cursor resting on preset number field (P##), enter desired preset number using keypad number key. Press ENT key.
    e. With cursor resting on type service field, press arrow keys to select CIR (Circuit). Press ENT key.
    f. With cursor resting on encryption field, press arrow keys to select UN (unencrypted). Press ENT key.
    g. If using data communications, press arrow keys to select SYN if the requested service is synchronous, or ASYN if asynchronous. Press ENT key.
    h. With cursor resting on V/D field, press arrow keys to select voice (V) or data (D) mode. Press ENT key.

NOTE: Will be using voice (V) mode.

    i. With cursor resting on data rate field, press arrow keys to select desired data rate. Press ENT key.
    j. With cursor resting on "Prec:" field, press arrow keys to select the precedence of the service connection. Press ENT key.
    k. With cursor resting on "D:#####" field, use keypad number keys to select the destination address to be called. Press ENT key.
    l. With cursor resting on "Code:###" field, use keypad number keys to enter the configuration code (000-511) for the requested service. Press ENT key.
    m. Press ESC once to return to SET PRESETS menu.

7. Load 5 kHz message presets
    a. Press the 4 key. (If at the MAIN MENU, press the 3 key to display SET PRESETS Menu, then press the 4 key.) The display shows the SET PRESET 5K MESG menu.
    b. With cursor resting on preset number field (P##), enter desired preset number using keypad number key. Press ENT key.
    c. With cursor resting on "TEK#" field, select desired Traffic Encryption Key (1-5). Press ENT key.
    d. With cursor resting on type encryption field, press arrow keys to select 4KG84. Press ENT key.
    e. With cursor resting on "Prec:" field, press arrow keys to select P (precedence of the service connection). Press ENT key.
    f. With cursor resting on encryption field, press arrow keys to select UN (unencrypted). Press ENT key.

STP 11-25C13-SM-TG

Performance Steps
g. With cursor resting on "D:#####" field, select destination address. (00000-65535). Press ENT key.
h. Press ESC once to return to SET PRESETS menu.

8. Load DAMA presets
   a. Press 1 key. (If at the MAIN MENU, press 3 key to display SET PRESETS Menu, then press 1 key.) The display shows the last used SET MODE PRESET menu.
   b. With cursor resting on mode field, press arrow keys to select DAMA. Press ENT key until cursor moves to next field.
   c. With cursor resting on preset number field (-P#), enter desired preset number using keypad number key. Press ENT key.
   d. With cursor resting on "TEK#" field, select desired Traffic Encryption Key (1-5). Press ENT key.
   e. With cursor resting on type encryption field, press arrow keys to select 4KG84. Press ENT key.
   f. With cursor resting on V/D field, press arrow keys to select voice (V) or data (D) mode. Press ENT key.

NOTE: Will be using voice mode.

   g. Perform this step only if using data mode. With cursor resting on data rate field, press arrow keys to select 2400. Press ENT key.
   h. With cursor resting on channel variant, press arrow keys to select 5 kHz. Press ENT key.
   i. With cursor resting on "Tpwr" field, press arrow keys to select power level of 38 dBm.
   j. With cursor resting on "Channel Number:###" field, enter desired operating channel number . Press ENT key. (The frequency corresponding to the entered channel number appears in the "R##.### T###.###" fields.)
   k. Enter configuration code. (Refer to TM 11-5820-1130-12&P, Appendix K.)
   l. With cursor resting on orderwire encryption field, press arrow keys to select OW:PT (orderwire encryption). Press ENT key.
   m. With cursor resting on mode of operation field, press arrow keys to select Normal. Press ENT key.
   n. With cursor resting on the ranging field, press arrow keys to display Passive. Press ENT key.
   o. With cursor resting on "Satellite ID:#" field, enter the ID of the satellite ephemeris (1-8) to be used for passive range. (Refer to TM 11-5820-1130-12&P, para 4.26.5 for satellite ID numbers.) Press ENT key.
   p. Press arrow keys to select Preassigned. Press ENT key. Also enter precedence with arrow keys and the demarcation with keypad number keys (00000-65536). Press ENT key after each field entry.

NOTE: The cursor moves back to the first data field (DAMA). This allows the operator to load additional presets.

9. Normal 5 kHz DAMA operation.
   a. Point the antenna in general direction of satellite.
   b. Verify DAMA is selected on CURRENT MODE menu. If not, press NEXT/PREV keys to move cursor to mode field. Press arrow keys until DAMA is displayed and then press ENT key.
   c. With cursor resting on preset number field (-P#), select desired operating preset using keypad number key (select the preset number even if number is already displayed). Press ENT key.
   d. Check for correct Traffic Encryption Key at "TEK#" field. Press ENT key.
   e. While observing field strength indication on display (Sq- -000), position satellite antenna for maximum field strength indication.
   f. Using NEXT/PREV keys, move cursor to "Start DAMA for #####" field on the CURRENT MODE menu and press ENT key. (The terminal will briefly display CONFIGURING WAIT on the CURRENT MODE menu. Then the display shows the DAMA operations menu indicating the operational mode (5 kHz) and the current terminal status.)
   g. When automatic login is complete, "Connected" appears on the display.

STP 11-25C13-SM-TG

Performance Steps

NOTE: If automatic login was unsuccessful, the LOGIN menu appears. In TM 11-5820-1130-12&P, paragraph 4.27 perform steps m and n.

10. Perform Circuit Service setup procedures.
    a. At DAMA operations menu, press hot key 1 to select "Service Setup."
    b. With cursor resting on service-preset field (P##), use keypad number keys to enter desired service preset. Press ENT key.
    c. With cursor resting on SEND, press ENT key.

NOTES:

(1) The terminal will return to the DAMA operations menu, displaying Pend on line 3 while channel controller processes your request for service setup. A popup message is displayed asking you to accept or reject the service setup with the parameters shown.

(2) If a WARNING or ERROR!! prompt is displayed on line 1, use NEXT/PREV key to scroll down and observe the message on line 5. Refer to TM 11-5820-1130-12&P, Table 4-16 and perform the necessary action listed for that message.

   d. With cursor resting on "TEK#" field, select desired key (1-5) with keypad number key. Press ENT key.
   e. With cursor resting on type encryption field, press arrow keys to select 4KG-84. Press ENT key.
   f. With cursor resting on ACCEPT/REJECT field, use arrow keys to select ACCEPT the service. Then press ENT key.
   g. The DAMA operations menu will indicate your service status on the third line.
   h. Transmit a message in voice operation. (Press and hold PTT switch on handset while talking in the mouthpiece (if in CT, wait for single beep in handset before transmitting).
   i. Receive a message in voice operation. (Release PTT switch and listen to the handset earpiece for squelch to break. Wait 10-15 seconds after receipt of transmission before pressing PTT.)

Evaluation Preparation: Setup: Ensure that all information, references, and equipment required to perform the task are available. Use the TMs and the evaluation guide to score the Soldier's performance.

Brief the Soldier. Tell the Soldier what he is required to do IAW the task conditions and standards.

Performance Measures	**GO**	**NO GO**
1. Prepared radio set for operation.	——	——
2. Performed power-up procedures.	——	——
3. Loaded keys into the radio set.	——	——
4. Loaded orderwire keys.	——	——
5. Configured terminal.	——	——
6. Loaded DAMA setups for 5 kHz service.	——	——
7. Loaded 5 kHz message presets.	——	——
8. Loaded DAMA presets.	——	——
9. Performed 5 kHz DAMA operation.	——	——
10. Sent/received communications using Voice mode of operation.	——	——

Evaluation Guidance: Score the Soldier a GO if all PMs are passed. Score the Soldier a NO-GO if any PM is failed. If the Soldier fails any PM, show what was done wrong and how to do it correctly. Have the Soldier perform the PMs until they are done correctly.

References

Required	**Related**
TM 11-5820-1130-12&P	

STP 11-25C13-SM-TG

Perform Scheduled Unit Level Maintenance (ULM) on Single-Channel Demand Assigned Multiple Access (DAMA) Tactical Satellite Terminal AN/PSC-5
113-589-3045

Conditions: Given radio set AN/PSC-5, digital multimeter AN/PSC-45, the ANCD with fill, fill cable, tool kit TK-101/G, TM 11-5820-1130-12&P, technical manual for the ANCD, AN/PSM-45, DA Form 2404, and DA Form 5988-E.

Standards: Documented all shortcomings on DA Form 2404 or DA Form 5988-E and corrected all deficiencies IAW TM 11-5820-1130-12&P; or evacuated the equipment that had a major deficiency beyond unit level maintenance to a higher maintenance level.

Performance Steps

1. Identify unit level maintenance (ULM) required. (Review DA Form 5988-E.)

2. Assemble required resources.

3. Perform ULM checks and services.

4. Take corrective actions.
    a. Repair correctable deficiencies or shortcomings.
    b. Evacuate the LRU with uncorrectable deficiencies to a higher maintenance level.
        (1) Record equipment status.
        (2) Report status to the ULLS clerk.
    c. Document uncorrected shortcomings.
        (1) Record actions taken.
        (2) Schedule next maintenance.

Performance Measures	GO	NO GO
1. Identified ULM required. (Reviewed DA Form 5988-E.)	——	——
2. Assembled required resources.	——	——
3. Performed ULM checks and services.	——	——
4. Took corrective actions.	——	——

    a. Repaired correctable deficiencies or shortcomings.
    b. Evacuated the LRU with uncorrectable deficiencies to a higher maintenance level.
        (1) Recorded equipment status.
        (2) Reported status to the ULLS clerk.
    c. Documented uncorrected shortcomings.
        (1) Recorded actions taken.
        (2) Scheduled next maintenance.

Evaluation Guidance: Score the Soldier a GO if all PMs are passed. Score the Soldier a NO-GO if any PM is failed. If the Soldier fails any PM, show what was done wrong and how to do it correctly. Have the Soldier perform the PMs until they are done correctly.

References
   Required
   ANCD TM
   DA FORM 2404
   DA FORM 5988-E
   TM 11-5820-1130-12&P

   Related
   DA PAM 738-750
   TM 11-6625-3052-14

STP 11-25C13-SM-TG

Operate Automated Net Control Device (ANCD) AN/CYZ-10
113-609-2053

Conditions: As a radio operator in a field environment, given data transfer device AN/CYZ-10 (C), TB 11-5820-890-12, TM 11-5820-890-10-8, DA Form 2404, and DA PAM 738-750. Given a requirement to operate the AN/CYZ-10.

Standards: Performed in sequence the transfer of COMSEC keys and SOI information from ANCD to ANCD; loaded SINCGARS radio with COMSEC variables using ANCD; obtained SOI information from ANCD, and corrected all errors within 15 minutes.

Performance Steps

1. Transfer COMSEC keys and SOI information from ANCD to ANCD.
    a. Turn on both ANCDs.
    b. Make main menu selection (ANCD).
    c. Make source ANCD menu selections.
    d. Make target ANCD menu selections.
    e. Transfer data from ANCD to ANCD.
    f. Turn OFF/disconnect ANCDs.

2. Load radio from ANCD using Mode 2 fill.
    a. Turn radio and ANCD power ON.
    b. Make main menu selection on ANCD
    c. Make application menu selection on ANCD
    d. Set controls of radio and connect to ANCD with fill cable.
    e. Transfer Mode 2 fill from ANCD to radio.
    f. Disconnect ANCD from radio and turn ANCD power OFF.

3. Obtain SOI information from ANCD
    a. Turn ANCD power ON.
    b. Make main menu selection.
    c. Make SOI menu selection
    d. Turn ANCD power OFF.

4. Perform PMCS on ANCD.
    a. Make a visual inspection of the ANCD.
    b. Check the battery.
    c. Check the fill port/CIK port.
    d. Record entries on DA Form 2404.

Performance Measures	**GO**	**NO GO**
1. Transferred COMSEC data and SOI information from ANCD to ANCD.	——	——
2. Loaded radio from ANCD using Mode 2 fill.	——	——
3. Obtained SOI information from the ANCD.	——	——
4. Performed PMCS on ANCD.	——	——

Evaluation Guidance: Score the Soldier a GO if all PMs are passed. Score the Soldier a NO-GO if any PM is failed. If the Soldier fails any PM, show what was done wrong and how to do it correctly. Have the Soldier perform the PMs until they are done correctly.

References
   Required   **Related**
   DA FORM 2404
   DA PAM 738-750
   TB 11-5820-890-12
   TM 11-5820-890-10-8

STP 11-25C13-SM-TG

Perform Net Control Station (NCS) Duties Using Automated Net Control Device (ANCD) AN/CYZ-10
113-609-2052

Conditions: Given a requirement and two Automated Net Control Devices (ANCDs), a SINCGARS radio set, operational radio net with two distant stations, W4 fill cable, TB 11-5820-890-10-12, and TM 11-5820-890-10-8.

Standards: Performed in sequence the transfer of COMSEC and SOI information from ANCD to ANCD; transmitted SOI data by Broadcast Mode; transmitted COMSEC Key to substations in a net by OTAR Manual Keying (MK) and OTAR Automatic Keying (AK) methods; and received and stored in an ANCD a COMSEC key sent by MK and OTAR, correcting all errors within 45 minutes.

NOTE: Supervision and assistance are available.

Performance Steps

1. Transfer partial COMSEC. (Refer to TM 11-5820-890-10-8, para 6.3c, pages 6-9 and 6-10.)
    a. Turn on both ANCDs.
    b. Make Main Menu selection on ANCD.
    c. Make Source ANCD Menu selections.
    d. Make Target ANCD Menu selections.
    e. Transfer data from ANCD to ANCD.
    f. Turn off/disconnect ANCDs.

2. Send SOI information by Broadcast mode. (Refer to TM 11-5820-890-10-8, Figure 4-19, page 4-32 and para 6.4d, pages 6-25–6-27.)
    a. Prepare NCS radio to send broadcast.
    b. Turn ANCD power ON.
    c. Make ANCD menu selection.
    d. Transfer COMSEC data (TEK) to NCS radio.
    e. Set controls of radio and connect to ANCD with fill cable.
    f. Send SOI information by Broadcast mode.
    g. Terminate task.

3. Send COMSEC data by MK over-the-air rekeying (OTAR). (Refer to TM 11-5820-890-10-8, Figure 4-15, page 4-29 and para 6.4e, pages 6-28–6-30.)
    a. Prepare source radio to send MK OTAR.
    b. Prepare source ANCD to send MK OTAR
    c. Connect ANCD to radio using fill cable.
    d. Make ANCD menu selection.
    e. Prepare target stations to receive MK OTAR.
    f. Send COMSEC data (TEK) by MK OTAR.

4. Receive TEK sent by MK OTAR. (Refer to TM 11-5820-890-10-8, para 4.16, page 4-29 and para 6.4g, pages 6-34 and 6-35.)

5. Store TEK sent by MK OTAR. (Refer to TM 11-5820-890-10-8, para 6.4g, pages 6-34 and 6-35.)

6. Send COMSEC data (TEK) by AK OTAR. (Refer to TM 11-5820-890-10-8, para 4.17, page 4-30 and para 6.4g, pages 6-31–6-33.)
    a. Prepare source radio for AK OTAR.
    b. Prepare ANCD to send AK OTAR
    c. Send COMSEC Key by AK OTAR
    d. Load TEK sent AK OTAR into source NCS radio.
    e. Update KEK used by AK OTAR in source radio.

Performance Measures	**GO**	**NO GO**
1. Transferred partial COMSEC from ANCD to ANCD.	——	——
2. Sent SOI information by Broadcast mode.	——	——
3. Sent COMSEC TEK by MK OTAR.	——	——
4. Received COMSEC TEK sent by MK OTAR.	——	——
5. Stored COMSEC TEK received by MK OTAR	——	——
6. Sent COMSEC TEK AK OTAR	——	——

Evaluation Guidance: Score the Soldier a GO if all PMs are passed. Score the Soldier a NO-GO if any PM is failed. If the Soldier fails any PM, show what was done wrong and how to do it correctly. Have the Soldier perform the PMs until they are done correctly.

References

    Required   **Related**
    TB 11-5820-890-12
    TM 11-5820-890-10-8

STP 11-25C13-SM-TG

Navigate Using the AN/PSN-11
113-610-2044

Conditions: This task is performed in a tactical or nontactical situation; given a requirement; operational precision lightweight global positioning system (GPS) receiver (PLGR) AN/PSN-11; battery, lithium, storage, BA-5800/U; COMSEC device KYK-13/TSEC or AN/CYZ-10; initialization, setup, and waypoint information; and TM 11-5825-291-13.

NOTE: Supervision and assistance is available.

Standards: Initialized and loaded the PLGR with COMSEC variables, entered setup and waypoint information, receiver acquired four satellites, and user successfully navigated to five waypoints within 60 minutes.

Performance Steps

1. Enter or verify correct setup displays.
    a. Select setup mode.
    b. Select setup units.
    c. Select setup magnetic variation type.
    d. Select elevation hold mode, time reference, and error display format.
    e. Select setup datum and automatic OFF timer.
    f. Select setup data port.
    g. Select setup automark.

2. Enter crypto keys.

3. Enter/verify initialization displays.
    a. Initialize position.
    b. Initialize time and date.
    c. Initialize track and ground speed.
    d. Initialize user-defined datum (page 1), if necessary.
    e. Initialize user-defined datum (page 2), if necessary.
    f. Initialize crypto key, if necessary.

4. Enter, edit, or review waypoint data.
    a. Copy waypoints.
    b. Determine distance between waypoints.
    c. Calculate a new waypoint.
    d. Clear waypoints.
    e. Define a mission route.

5. Check status displays to ensure acquisition of four satellites

6. Navigate to five waypoints.

Evaluation Preparation: Setup: For this evaluation, prepare equipment and ensure equipment is operational. Provide sufficient initialization, setup, and waypoint data information to the Soldier so that the PLGR can be used to navigate. Prepare a navigation course.

Brief Soldier: Tell the Soldier all PMs must be accomplished within 60 minutes.

Performance Measures (Refer to TM 11-5825-291-13 for all PMs.)	GO	NO GO
1. Entered or verified correct setup displays.	——	——
2. Entered crypto keys.	——	——

STP 11-25C13-SM-TG

Performance Measures	**GO**	**NO GO**
3. Entered and verified initialization displays.	——	——
4. Entered waypoint information.	——	——
5. Checked status displays to ensure acquisition of four satellites.	——	——
6. Navigated to five waypoints.	——	——

Evaluation Guidance: Score the Soldier a GO if all PMs are passed. Score the Soldier a NO-GO if any PM is failed. If the Soldier fails any PM, show what was done wrong and how to do it correctly. Have the Soldier perform the PMs until they are done correctly.

References
    Required   **Related**
    TM 11-5825-291-13

STP 11-25C13-SM-TG

Subject Area 7: AUTOMATION OPERATIONS

Configure a Desktop IBM or Compatible Microcomputer for Operation
113-580-1032

Conditions: Given a central processing unit (CPU), monitor, keyboard, mouse, mouse pad, printer, surge suppressor, Microsoft-Disk Operating System (MS-DOS), Windows Operating System (OS), and application software.

Standards: Displayed and printed text using Notepad or WordPad.

Performance Steps

1. Prepare for configuration.
    a. Inventory components from enclosed packing lists.
    b. Select appropriate site.
        (1) Enough desktop space for computer and any peripheral devices.
        (2) Close proximity to electrical connection.

2. Configure the microcomputer.
    a. Place the CPU in the selected location.
        (1) Plug the female end of the power cable into the connector on the rear panel of the CPU.
        (2) Plug the male end into the surge suppressor.
    b. Place the keyboard in front of the CPU.
        (1) Plug the keyboard cable into the keyboard connector on the rear panel of the CPU.
        (2) Adjust the keyboard legs as desired.
    c. Connect the monitor.
        (1) Place the monitor on or near the CPU.
        (2) Connect the monitor power cable to the surge suppressor.
        (3) Connect the monitor signal cable to the monitor connector on the rear panel of the CPU.
        (4) Tighten the two screws on the monitor signal cable plug to anchor the cable to the connector.
    d. Connect printer.
        (1) Connect printer interface cable to the printer connector on the rear panel of the printer.
        (2) Connect the opposite end of the printer interface cable to a printer connector on the back panel of the CPU.
        (3) Tighten the two screws of the printer interface cable to anchor the cable to the connector.
        (4) Connect the printer power cable to the surge suppressor.
        (5) Set up printer paper.
    e. Connect mouse.
        (1) Connect the mouse to the appropriate serial port.
        (2) Tighten the two screws on the cable plug to anchor the cable to the connector.
        (3) Place the mouse pad next to the keyboard and set the mouse on it.
    f. Connect AC power.
        (1) Ensure the system power switch and surge suppressor switches are set to OFF.
        (2) Ensure that the CPU line voltage selection switch is set to the proper voltage.
        (3) Plug surge suppressor into the AC outlet.

3. Set up the system.
    a. Apply power to the system.
    b. If no operating system was factory installed, install the operating system.
    c. Install application software.

4. Test the system.
    a. Create text in Notepad or WordPad.
    b. Save and print text.

Performance Measures	GO	NO GO
1. Prepared for configuration.	___	___
2. Configured the microcomputer.	___	___
3. Set up the system.	___	___
4. Tested the system.	___	___

Evaluation Guidance: Score the Soldier a GO if all PMs are passed. Score the Soldier a NO-GO if any PM is failed. If the Soldier fails any PM, show what was done wrong and how to do it correctly. Have the Soldier perform the PMs until they are done correctly.

References

Required	**Related**
DOS MANUAL	CUM
WUM	MOM
	PUM

STP 11-25C13-SM-TG

Install Network Hardware/Software in a Desktop IBM or Compatible Microcomputer
113-580-1033

Conditions: Given an operational IBM or compatible microcomputer, network interface card, network cable, T-connector, terminator, electrostatic discharge (ESD) wrist strap, network software, local area network (LAN), cable splicing kit, tool kit TK-101/G, Network Software User's Manual (NSUM), and Network Card Installation Manual (NCIM).

Standards: Entered the computer into the network, sent a test E-mail, and received a reply.

Performance Steps

1. Prepare for installation.

CAUTION: Circuit card assemblies are sensitive to ESD. Wear an ESD wrist strap whenever the CPU cover is removed.

    a. Collect all required materials.
    b. Quit all applications and shutdown the system.
    c. Turn power off and unplug power cord.
    d. Remove the CPU cover.
    e. Select an unused expansion slot.
    f. Determine bus expansion slot compatibility with the network card.
        (1) Industry standard architecture (ISA), 8- or 16-bit bus.
        (2) Peripheral component interconnect (PCI), 32- or 64 bit-bus.

2. Configure a network card. (Refer to the NCIM to determine settings.)

3. Install a network card.
    a. Remove the cover from the rear slot and save the screw for a later use.

CAUTION: Do not touch the gold contacts on the bottom of the network card.

    b. Remove the network card from its protective wrapping.
    c. Align the bottom of the card (gold contacts) with the selected slot and press down firmly.
    d. Anchor the card in place with the screw removed in step 3a above.
    e. Replace the CPU cover.
    f. Restore power to the CPU.

4. Connect the computer to the network. (Refer to the NCIM to determine cable installation.)

5. Install network software. (Refer to the NSUM for installation instructions.)

6. Create an E-mail account. (Refer to the NSUM.)

7. Send an E-mail message. (Refer to the NSUM.)

8. Receive an E-mail message. (Refer to the NSUM.)

Performance Measures   <u>GO</u>   <u>NO GO</u>

1. Prepared for installation.

CAUTION: Circuit card assemblies are sensitive to ESD. Wear an ESD wrist strap whenever the CPU cover is removed.

    a. Collected all required materials.
    b. Quit all applications and shut down the system.
    c. Turned off power and unplugged the power cord.
    d. Removed the CPU cover.
    e. Selected an unused bus expansion slot.

Performance Measures	<u>GO</u>	<u>NO GO</u>

    f. Determined bus expansion slot compatibility with the network card (see Figure 3-1).
        (1) ISA, 8- or 16-bit bus.
        (2) PCI, 32- or 64-bit bus.

2. Configured a network card. (Refer to the NCIM to determine settings.)   ____   ____

3. Installed the network card.   ____   ____
    a. Removed the cover from the rear slot and saved the screw for later use.

CAUTION: Do not touch the gold contacts on the bottom of the network card.

    b. Removed the network card from its protective wrapping.
    c. Aligned the bottom of the card (gold contacts) with the selected slot and pressed down firmly.
    d. Anchored the card in place with the screw removed in step 3a.
    e. Replaced the CPU cover.
    f. Restored power to the CPU.

4. Connected the computer to the network. (Refer to the NCIM to determine cable installation.)   ____   ____

5. Installed network software. (Refer to the NSUM for installation instructions.)   ____   ____

6. Created an E-mail account. (Refer to the NSUM.)   ____   ____

7. Sent an E-mail message. (Refer to the NSUM.)   ____   ____

8. Received an E-mail message. (Refer to the NSUM.)   ____   ____

Evaluation Guidance: Score the Soldier a GO if all PMs are passed. Score the Soldier a NO-GO if any PM is failed. If the Soldier fails any PM, show what was done wrong and how to do it correctly. Have the Soldier perform the PMs until they are done correctly.

References

Required	**Related**
NCIM	CUM
NSUM	WUM

STP 11-25C13-SM-TG

Install a Tactical Local Area Network (LAN)
113-580-1035

Conditions: Given a network capable PC compatible or equivalent with manuals, appropriate software with user's manual (UM), cables connectors, hubs, SOI, and site map.

Standards: The tactical local area network was able to send and receive a test message.

Performance Steps

1. Review site map.
    a. Identify network type.
    b. Identify assigned Internet protocol (IP) addressing.
    c. Identify required resources.

2. Inventory resources.

3. Connect hardware/peripherals to network.
    a. Place stipulated equipment IAW the site map.
    b. Connect all hardware/peripherals IAW the site map.

4. Initialize PC systems.
    a. Power up hardware.
    b. Verify functionality of PC systems.
    c. Ensure that appropriate OS and network software is installed.

5. Configure network software.
    a. Assign IP address to each system as appropriate.
    b. Set up user profiles IAW the unit's SOI/SOP.
    c. Test connectivity (that is ping).

6. Perform operational check.
    a. Send a test message to all addressees.
    b. Receive an acknowledgement from all addressees.

Evaluation Preparation: Setup: Ensure adequate resources are available and a site map and SOI are provided.

Brief Soldier: NA.

Performance Measures	**GO**	**NO GO**

1. Reviewed site map.
     a. Identified network type.
     b. Identified assigned IP addressing.
     c. Identified required resources.
	——	——

2. Inventoried resources.	——	——

3. Connected hardware/peripherals to network.
     a. Placed stipulated equipment IAW the site map.
     b. Connected all hardware/peripherals IAW the site map.
	——	——

4. Initialized PC systems.
     a. Powered up hardware.
     b. Verified functionality of PC systems.
     c. Ensured that appropriate OS and network software was installed.
	——	——

5. Configured Network Software.	——	——

Performance Measures   <u>GO</u>   <u>NO GO</u>

    a. Assigned IP address to each system as appropriate.
    b. Set up user profiles IAW the unit's SOI/SOP.
    c. Tested connectivity (that is ping).

  6. Performed operational check.   ___   ___

NOTE: If the operational check failed, refer to tasks 113-580-0053, 113-580-0052, 113,580-0040, and 113-580-0051 in this manual as applicable.

    a. Sent a test message to all addressees.
    b. Received an acknowledgement from all addressees.

NOTE: If the operational check failed, refer to tasks 113-580-0051, 113-580-0052, 113-580-0053, and 113-580-0055 in this manual as applicable.

Evaluation Guidance: Score the Soldier a GO if all PMs are passed. Score the Soldier a NO-GO if any PM is failed. If the Soldier fails any PM, show what was done wrong and how to do it correctly. Have the Soldier perform the PMs until they are done correctly.

References
    Required   **Related**
    MFG MANUALS
    SUM
    UNIT SOI
    UNIT SOP

STP 11-25C13-SM-TG

Subject Area 8: REMOTE COMMUNICATIONS

Install Radio Set Control Group AN/GRA-39(*)
113-622-1006

Conditions: Given a tactical or nontactical situation; a requirement; radio set control group AN/GRA-39(*); AN/VRC-12 series or AN/PRC-25/77 series FM radio set, installed and operational; batteries BA-30, 12 each; field wire on reels (up to 2 miles); tool equipment TE-33; designated remote site for the radio; vehicle driver/radio operator; operating frequency from unit SOI; initiated DA Form 2404 or DA Form 5988-E; TM 11-5820-477-12, and TM 38-750. This task may be performed in an NBC environment.

Standards: Prepared, positioned at the designated locations, and connected the AN/GRA-39(*) according to the PMs.

Performance Steps

1. Install batteries in the AN/GRA-39(*). (Refer to TM 11-5820-477-12.)
    a. Unsnap the clamps holding the rear covers of the case and remove the covers. (Refer TM 11-5820-477-12, Figure 3-10.)
    b. Inspect the battery compartments.
        (1) Clean the battery compartments and the contact strips. (A pencil can be used to clean the contacts.)
        (2) Check the battery compartments for cracks.
    c. Install six batteries in each unit. Observe the polarity indicated on the battery box. (Refer TM 11-5820-477-12, Figure 3-11.)
        (1) Ensure all contacts touch the battery terminals.
        (2) Ensure the batteries are held in place under tension.
    d. Replace covers and snap the clamps into place.
        (1) Position the covers squarely on the cases, without sliding them.
        (2) Use both hands to secure both clamps of each unit at the same time.

NOTE: If the cover has slid into position or the clamps are closed one at a time, the batteries can be pushed away from the contacts. Batteries in the local control unit should be replaced after approximately 72 hours of operation. Batteries in the remote control unit should be replaced after approximately 24 hours of operation. Batteries must be removed from the equipment on days the equipment is not being used.

2. Install control units on site. (Refer to TM 11-5820-477-12, Figure 3-12.)
    a. Local control C-2329(*)/GRA-39.
        (1) Next to or on the receiver-transmitter.
        (2) Connect the connector of the radio cable to the receptacle of the radio receiver-transmitter.
    b. Remote control C-2328(*)/GRA-39 at the remote radio operation site (up to 2 miles from the radio/local control unit).

3. Install field wire line between the sites.

NOTE: Tie the field wire to a solid object prior to connecting to the control unit binding posts. This will prevent damage to the control units if the wire should be snagged and pulled.

   a. Remove ½ inch of insulation from each conductor of the field wire, press down one binding post, insert the bared end of one wire into the slot, and release the binding post. Repeat for the other binding post.
   b. Lay the field wire to the remote control site (up to 2 miles).

STP 11-25C13-SM-TG

Performance Steps

WARNING: Do NOT press the RINGER button of either the local or remote control unit while installing the field wire to the line binding posts. Operating the RINGER button applies a voltage, which can shock a person touching the binding posts or the bare conductors.

    c. Repeat performance step 3a for the remote control unit.

4. Update DA Form 2404 or DA Form 5988-E. (Refer to TM 38-750 and TM 11-5820-477-12.)

NOTE: To dismount the radio set, reverse the procedures outlined in performance steps 1 through 3, observing all applicable WARNINGS, CAUTIONS, and NOTES.

Performance Measures	<u>GO</u>	<u>NO GO</u>
1. Installed batteries in the AN/GRA-39(*). (Refer to TM 11-5820-477-12.)	——	——
2. Installed control units on site. (Refer to Figure 3-12 and TM 11-5820-477-12.)	——	——
3. Installed field wire line between the sites.	——	——
4. Updated DA Form 2404 or DA Form 5988-E. (Refer to TM 38-750 and TM 11-5820-477-12.)	——	——

Evaluation Guidance: Score the Soldier a GO if all PMs are passed. Score the Soldier a NO-GO if any PM is failed. If the Soldier fails any PM, show what was done wrong and how to do it correctly. Have the Soldier perform the PMs until they are done correctly.

References
    Required   **Related**
    DA FORM 2404
    DA FORM 5988-E
    TM 11-5820-477-12
    TM 38-750
    UNIT SOI

STP 11-25C13-SM-TG

Operate Radio Set Control Group AN/GRA-39(*)
113-622-2004

Conditions: Given a tactical or nontactical situation; a requirement; installed radio set control group AN/GRA-39(*), AN/GRC-193A, AN/PRC-1048(V)3, AN/GRC-213A, and AN/GRC-193B, installed and operational; another radio station with which to communicate; operator/attendant for the radio site; TM 11-5820-924-13, and TM 11-5820-477-12. This task may be performed in an NBC environment.

Standards: Performed starting procedures and communication checks within 10 minutes.

Performance Steps

1. Perform starting procedures.

NOTE: Starting procedures for the local control and radio are performed by the radio operator/attendant. The remote operator performs starting procedures for the remote control. Starting procedures can be coordinated through the telephone feature of the AN/GRA-39(*).

    a. Local control C-2329(*)/GRA-39. (Refer to TM 11-5820-477-12, Figure 3-13.)
        (1) Turn the POWER switch to ON.
        (2) Turn the BUZZER VOLUME control to approximately midrange.
        (3) Connect a handset to the audio connector of the local control.

NOTE: The AN/GRA-39(*) has one component handset H-189/GR. In order for the local control operator to use the telephone feature, this handset is required. The remote control operator, in conjunction with the RAD/SPKR switch, may use the microphone from the radio set.

        (4) Set the VOLUME control of the radio set for a comfortable listening level in the handset.

NOTE: If the radio set tries to self-key (or chatter), lower the VOLUME level of the receiver-transmitter. If local control C-2329(*)/GRA-39 or C-2329A/GRA-39 has not had MWO 11-5820-477-30/1 applied and an RT-841/PRC-77 or RT-505/PRC-225 is used as the radio set, the FUNCTION switch of the RT cannot be set to SQUELCH. Additionally, if the RT-505/PRC-25 is used, the other stations in the net must also operate without squelch, or in the OLD SQUELCH mode.

    b. Remote control C-2328(*)/GRA-39. (Refer to TM 11-5820-477-12, Figure 3-14.)
        (1) Turn the VOLUME control to approximately midrange.
        (2) Turn the BUZZER VOLUME control to approximately midrange.
        (3) Connect a handset to the audio connector.
        (4) Set the TEL-RAD-RAD/SPKR switch to the TEL position.

2. Conduct a telephone communications check between local and remote control units. (Refer to TM 11-5820-477-12.)

WARNING: Do not touch the LINE binding posts or the bare portions of the field wire while pumping the RINGER button.

    a. Pump the RINGER button several times in quick succession to gain the attention of the other operator.
    b. Local control C-2329(*)/GRA-39. Turn and hold the TEL-REMOTE-RADIO switch to TEL while talking and listening.

NOTE: If microphone M-80(*)/U is used at the remote control, the FUNCTION switch must be set to TEL to talk and to RAD/SPKR to listen to the other operator.

    c. Press the handset PTT switch to talk to the operator; release to listen.

3. Conduct a radio transmission and reception check. (Refer to TM 11-5820-477-12.)
    a. Local control C-2329(*)/GRA-39.
        (1) Turn and hold the TEL-REMOTE-RADIO switch to RADIO.

STP 11-25C13-SM-TG

Performance Steps
    (2) Press the handset PTT switch to transmit; release to receive. Use proper radio call signs and procedures.
   b. Remote control C-2328(*)/GRA-39.
    (1) Set the TEL-RAD-RAD/SPKR switch to either RAD or RAD/SPKR.
    (2) Adjust the VOLUME control to the desired listening level in the handset or loudspeaker.
    (3) Press the handset PTT switch to transmit; release to receive. Use proper radio call signs and procedures.

NOTE: The TEL-REMOTE-RADIO switch on the local control must be in the REMOTE position during this operation.

  4. Stopping procedures. (Refer to TM 11-5820-477-12.)
   a. Local control, turn the POWER switch to OFF.
   b. Remote control, turn the VOLUME control to OFF.

Performance Measures   <u>GO</u>   <u>NO GO</u>

1. Performed starting procedures.   ——   ——

2. Conducted a telephone communications check between local and remote control units. (Refer to TM 11-5820-477-12.)   ——   ——

3. Conducted a radio transmission and reception check. (Refer to TM 11-5820-477-12.)   ——   ——

4. Performed stopping procedures. (Refer to TM 11-5820-477-12.)   ——   ——

Evaluation Guidance: Score the Soldier a GO if all PMs are passed. Score the Soldier a NO-GO if any PM is failed. If the Soldier fails any PM, show what was done wrong and how to do it correctly. Have the Soldier perform the PMs until they are done correctly.

References
  Required   **Related**
  TM 11-5820-477-12
  TM 11-5820-924-13

STP 11-25C13-SM-TG

Perform Unit Level Maintenance on Radio Set Control Group AN/GRA-39(*)
113-622-3001

Conditions: As a Radio Operator-Maintainer in a field environment, given a tactical or nontactical situation, Radio Set Control Group AN/GRA-39, tool kit TK -101, lint-free cloth, mild detergent solution, TM 11-5820-477-12, DA PAM 738-750, and DA Form 2404 or DA Form 5988-E.

NOTE: Supervision and assistance are available.

Standards: Cleaned the exterior of the radio set control group, tightened all dials and knobs; recorded all uncorrectable faults on DA Form 2404 or DA Form 5988-E without error, and reported to immediate supervisor according to PMs.

Performance Steps

1. Perform PMCS on radio set control group AN/GRA-39. (Refer to TM 11-5820-477-12.)

WARNING: Trichlorotrifluoroethane is DANGEROUS. Adequate ventilation should be provided while using it. Prolonged breathing of its vapor should be avoided. The solvent should not be used near heat or an open flame: the products of decomposition are toxic and irritating. Since trichlorotrifluoroethane dissolves natural oils, prolonged contact with skin should be avoided. When necessary, use gloves, which the solvent cannot penetrate. If the solvent is taken internally, consult a physician immediately.

2. Correct defects if necessary. Replacement parts or materials are obtained from your team chief. (Refer to TM 11-5820-477-12.)

3. Complete DA Form 2404 or DA Form 5988-E as a daily maintenance form. (Refer to DA PAM 738-750.)

4. Report all uncorrectable defects. (Refer to DA PAM 738-750.)
    a. Notify immediate supervisor of all uncorrectable faults found.
    b. Submit DA Form 2404 or DA Form 5988-E to supervisor or maintenance support personnel.

Evaluation Preparation: Setup: Ensure equipment has preplanned defects.

Brief Soldier: Tell the Soldier all PMs must be passed in sequence.

Performance Measures	GO	NO GO
1. Performed PMCS on radio set control group.	——	——
2. Corrected defects if necessary. Replaced parts or materials obtained from your team chief.	——	——
3. Completed DA Form 2404 or DA Form 5988-E as a daily maintenance form.	——	——
4. Reported all uncorrectable defects.	——	——

Evaluation Guidance: Score the Soldier a GO if all PMs are passed. Score the Soldier a NO-GO if any PM is failed. If the Soldier fails any PM, show what was done wrong and how to do it correctly. Have the Soldier perform the PMs until they are done correctly.

References
- **Required**
  - DA FORM 2404
  - DA FORM 5988-E
  - DA PAM 738-750
  - TM 11-5820-477-12
- **Related**

STP 11-25C13-SM-TG

Subject Area 15: ABCS CORE

Install Force XXI Battle Command Brigade and Below (FBCB2)
113-580-1055

Conditions: Given FBCB2 hardware, DTM 11-7010-326-10, DTM 11-7010-326 -20&P, installation kit, multimeter AN/PSM-45(*), TK-101/G, SB 11-131-2, DA PAM 738-750, Contract Data Requirements List GO17 FBCB2 Unit Maintenance Manual (CDRL GO17 FBCB2 UMM), and DA Form 5988-E.

Standards: Installed the FBCB2 IAW SB 11-131-2 and sent and received a test message.

Evaluation Preparation: Ensure adequate resources are available.

Performance Measures   <u>GO</u>   <u>NO GO</u>

1. Reviewed the tactical Internet (TI) chart.
   a. Identified system type.
   b. Identified assigned unit requirement number (URN).
   c. Identified required resources.

2. Inventoried installation kit. (Refer to SB 11-131-2.)

3. Mounted system components IAW CDRL GO17 FBCB2 UMM.
   a. Placed stipulated equipment.
   b. Connected all hardware/peripherals.

4. Initialized the systems.
   a. Powered up the AN/PSN-11 Precision Lightweight Global Positioning System Receiver (PLGR).
   b. Powered up the Enhanced Position Location Reporting System (EPLRS) if installed.
   c. Powered up the Single-Channel Ground and Airborne Radio System (SINCGARS) radio sets.
   d. Verified device operation.
   e. Powered up the AN/UYK-128 appliqué.

5. Configured the appliqué with the URN.
   a. Entered system administration from the session manager.
   b. Inputted the URN from the TI chart.

6. Performed an operational check.
   a. Sent a test message to all addressees.
   b. Received an acknowledgment from all addressees.

7. Established maintenance records IAW DA PAM 738-750.

Evaluation Guidance: Score the Soldier a GO if all PMs are passed. Score the Soldier a NO-GO if any PM is failed. If the Soldier fails any PM, show what was done wrong and how to do it correctly. Have the Soldier perform the PMs until they are done correctly.

References
    Required
    CDRL GO17 FBCB2 UMM
    DA FORM 5988-E
    DA PAM 738-750
    DTM 11-7010-326-10
    DTM 11-7010-326-20&P
    SB 11-131-2

    Related
    GPS-DC MK III OSM
    TSP 113-587-1067

STP 11-25C13-SM-TG

PREPARE/SEND COMBAT MESSAGES USING FBCB2 VERSION 3.4
171-147-0001

Conditions: Given a vehicle with an operational AN/UYK-128 Force XXI Battle Command Brigade and Below (FBCB2) system with software version 3.4 loaded, SINCGARS, GPS, and EPLRS (if equipped), and tactical situation requiring you to prepare and send combat messages. Default Message Addressing settings have already been set.

Standards: As a minimum, you must prepare and send one of each of the Combat Messages (SALT, MEDEVAC, NBC 1, FIRE MISSION, CHECK FIRE ALL, and SITREP) using FBCB2.

Performance Steps

1. Select the Combat Msgs button from the Function bar or the F3 button on the keyboard. The system will display the Combat Messages dialog box.

NOTE: Combat Messages are designed to quickly create and send mission essential messages during combat operations. Combat Messages are easier and require fewer steps than Long Form Messages. Use the Touch screen or embedded pointing device to make the entries. Combat Messages do not require keystrokes to make the entries. Combat Messages with programmed transmission settings provide the capability to build and send a message within 10 seconds. The Combat Messages function has six tabs which include: SALT = Size, Activity, Location, and Time; MEDEVAC = Medical Evacuation; NBC1 = Nuclear, Biological, Chemical; FIRE MISSION; CHECK FIRE ALL; SITREP = Situations Report. The system defaults to the SALT tab or the last tab selected when using Combat Messages.

2. Prepare and send a SALT message.

NOTE: Subordinate units use the SALT message to provide combat intelligence information on enemy units to higher echelons. The user can initiate a call-for-fire message from within this template. The SALT message automatically generates and broadcasts enemy georeference icons to friendly platforms throughout the Brigade. The user can submit an abbreviated Call for Fire Message using the information entered in the SALT Message. The dual message capability saves the user time and eliminates repetitive data input.

   a. Select the SALT tab. The system displays the combat messages SALT template.
   b. Select the equipment pull-down arrow. The system will display the equipment options list.

NOTE: If there is more than one different type of equipment, put the most dangerous target in the first equipment text box.

   c. Select an option from the list. The system displays the selection in the equipment combo box.
   d. Select the + or - button to the right of the Equipment pull-down arrow. Select the + button to increase the number or the - button to decrease the number.

NOTE: There are two additional Equipment Combo boxes if there is more than one type of equipment that is observed.

   e. Select the activity pull-down arrow. The system will display the activity options list.
   f. Select an option from the list. The system displays the selection in the activity combo box.
   g. Select the Map button. The Combat Messages dialog box will disappear and the mouse cursor will be replaced with the cross hair.
   h. Select a location on the map where the target is located. The Combat Messages dialog box will reappear with the location in the location text box or select LRF if the user's platform has a Laser Range Finder and Lase the target. The location will appear in the Location text box.
   i. Select the NOW button or the DTG button to update the Date Time Group.

Performance Steps

NOTE: The Date Time Group auto-fills when the user opens a message but does not update automatically. Selecting the NOW button enters the current DTG. Selecting the DTG button makes the user manually enter the DTG.

    j. Select the Speed pull-down arrow. The system will display the Speed options list.

NOTE: If you picked stationary for the Activity field, then you would select None for speed.

    k. Make a selection from the list. The system will display the selection in the Speed text box.
    l. Select the Course pull-down arrow. The system will display the Course option list.
    m. Select the direction that the equipment is moving or oriented in. The system will display the selection in the Course text box.
    n. Select the CFF Msgs Check box. The system will display a check mark in the box, which will give access to the Method of Engagement window.

NOTE: This function enables a request for fire support when sending a SALT message.

    o. Select the Method of Engagement combo box down arrow button. The system will display the Fire Request Type options list.
    p. Select an option from the list. The system displays the selection in the Method of Engagement text box.
    q. Select the Save button to save the message. The system will save the message and display Message saved to file Spot with the DTG after it.
    r. Select the Send button to send the message. The system will transmit the message and display "Last SALT message sent at DTG."
    s. Select the Close button to close the Combat Messages dialog box. The system will close the combat messages dialog box.

3. Prepare and send a fire mission message.
    a. Select the Combat Msgs button on the Function Bar or the F3 button on the keyboard.
    b. Select the Fire Mission tab. The system displays the Fire Mission tab group.

NOTE: The Fire Mission template is used to request indirect fire support from supporting fire support units. The Fire Mission Tab group consists of: Summary, CFF, Subsequent Adjust, Check Fire, On Call Fire Cmd, and EOM tabs. The Summary and CFF tabs are enabled at all times. The Subsequent Adjust, Check Fire, On Call Fire Command, and End of Mission tabs are grayed out. These tabs only become active when a Fire Mission is approved by the Advanced Field Artillery Tactical Data System (AFATDS).

    c. Select the CFF tab to access the Call For Fire message template.
    d. Select the Type of Mission pull-down arrow. Select an option from the list. The system will display the Type of Mission in the Type of Mission text box.
    e. Select the equipment pull-down arrow. The system will display the equipment options list.
    f. Select an option from the list. The system displays the selection in the Equipment text box.
    g. Enter the quantity by selecting the + or - button for the amount of equipment that the user observes or select the text box and enter the amount with the keyboard. The system will display the amount in the text box.
    h. Select the Map button. The Combat Messages dialog box will disappear and the mouse cursor will be replaced with the cross hair.
    i. Select the location on the map where the target is observed. The Combat Messages dialog box reappears and displays the location in the Target Location text box.
    j. Select the Fill LRF button. The system will input the location provided by the laser range finder (LRF).

NOTE: The LRF button only works with vehicles equipped with a Laser Range Finder.

    k. Select the Protection Level pull-down arrow. The system will display the Protection Level options list. Select an option from the list. The system displays the selection in the Protection Level text box.

STP 11-25C13-SM-TG

Performance Steps

NOTE: The Protection Level combo box is grayed out and only becomes active when Dismounted Personnel is selected in the Equipment combo box.

 l. Select the Method of Control pull-down arrow. The system will display the Method of Control options list. Select an option from the list. The system displays the selection in the Method of Control text box.

NOTE: The Timed Time On Target (Zulu) field only becomes active if the Timed Time on Target is selected in the Method Of Control field.

 m. Select the Save button if you desire to save the message. The system will save the message.
 n. Select the Send button if you desire to send the message. The system will transmit the message.
 o. Select the Summary tab, to display the Summary tab group.

NOTE: The Summary tab displays a list of all fire missions received along with their current status. You can access additional fire support messages from the Summary tab group after a Call For Fire message has been sent and a Message to Observer (read only) message has been received.

 p. Select the Close button. The system will return to the Ops screen.

4. Prepare and send a medical evacuation (MEDEVAC) request.

NOTE: Use the MEDEVAC request message to request ground or aircraft support to evacuate friendly and/or enemy casualties. The MEDAVAC tab group is accessed from Combat Msgs button on the Function Bar or the F3 button on the keyboard.

 a. Select the Combat Msgs button on the Function Bar or the F3 button on the keyboard.
 b. Select the MEDEVAC tab. The system will display the MEDEVAC template.
 c. Select the Map button. The Combat Messages dialog box will disappear and the mouse cursor will be replaced with a cross hair.
 d. Select a location on the map and the grid location will appear in the Pick Up Location text box.
 e. Or Select the LRF button. The system will input the location provided by the Laser Range Finder.

NOTE: This function only works if your vehicle is equipped with the Laser Range finder.

 f. Select the Amb Patients + or - buttons to input the amount of Amb patients or place the cursor in the Amb Patients text box and enter the amount with the keyboard.

NOTE: Holding down on the + or - button will scroll through the numbers at a fast pace.

 g. Select the Ltr Patients + or - buttons to input the amount of Litter patients or place the cursor in the Litter Patients text box and enter the amount with the keyboard.
 h. Select the Marking pull-down arrow. The system will display the Marking options list.
 i. Select an option from the list. The system displays the selection in the Marking text box.
 j. Select the Color pull-down arrow. The system will display the Color options list.
 k. Select an option from the list. The system displays the selection in the Color text box.
 l. Select the Pickup Zone Hot check box. The system will display a check mark in the Pickup Zone Hot text box.

NOTE: If the area is not Hot, then do not check this box. Hot means that the area has hostiles in the vicinity.

 m. Select the MEDAVAC Priority pull-down arrow. The system will display the MEDAVAC Priority options list.
 n. Select and option from the list. The system will display the selection in the MEDEVAC Priority text box.
 o. Select the NBC Contamination pull-down arrow if there is any type of contamination in the area. The system will display the NBC Contamination option list.

3 - 114   16 February 2005

Performance Steps

    p. Make a selection from the option list. The selection will display in the NBC Contamination Type text box.
    q. Select the Security pull-down arrow. The system will display the Security option list.
    r. Make a selection from the option list. The selection will display in the Security text box.
    s. Select the Save button if you desire to save the message. The system will save the message and display the Message saved to file spot DTG.
    t. Select the Send button if you desire to send the message. The system will display the Last MEDEVAC message sent with the DTG.
    u. Select the Close button when you desire to close the message box. The system will close the Combat Messages dialog box.

5. Prepare and send a nuclear, biological, chemical (NBC 1) message.

NOTE: NBC 1 is accessed from Combat Msgs button on the Function Bar or the F3 button on the keyboard. The Nuclear, Biological, Chemical (NBC 1) message allows the user to transmit observer's basic data on a single NBC attack. The message will automatically create a georeference location on the SA Display.

    a. Select the Combat Msgs. button on the Function Bar or the F3 button on the keyboard.
    b. Select the NBC 1 tab. The system will display the NBC 1 template.
    c. Select the NBC Event Type pull-down arrow. The system will display the NBC Event Type options list.
    d. Select an option from the list. The system displays the selection in NBC Event Type text box.
    e. Select the Delivery Means pull-down arrow. The system will display the Delivery Means option list.
    f. Make a selection from the option list. The system will display the selection in the text box.
    g. Select the first Map button to the right of the Attack Location 1 text box. The Combat Messages dialog box will disappear and the cursor will be replaced with a cross hair.

NOTE: There are two Attack Location fields. Do the same for the second Attack Location.

    h. Select a location on the map and the system will display the location in the Attack Location 1 text box.
    i. Or select the LRF button and lase the location. The location will appear in the Attack Location 1 text box.
    j. Enter the Attack Time 1(Zulu) data by selecting the NOW button Or, Selecting the plus + or - buttons to increase or decrease the Day, Hour and Minute.
    k. Enter the Attack Time 2(Zulu) data by selecting the NOW button Or, Selecting the plus + or - buttons to increase or decrease the Day, Hour and Minute.
    l. Select the Save button if you desire to save the message.
    m. Select the Send button if you desire to send the message. The system will transmit the message.
    n. Select the Close button. The system will close the Combat Messages dialog box.

6. Prepare and send a Check Fire message.
    a. Select the Combat Msgs. button on the Function Bar or the F3 button on the keyboard.
    b. Select the Check Fire All tab. The system will display the Check Fire All template.

NOTES:

(1) The Check Fire All tab allows the user to send an immediate Check Fire command for all active fire missions in the Summary tab group.

(2) To cancel a Check Fire All, just select the Cancel Check Fire All radio button.

(3) The Cancel Check Fire All radio button will be grayed out until the Check Fire All radio button is selected.

Performance Steps

    c. Select the Check Fire All radio button. The system will Highlight the Check Fire All radio button.
    d. Select the Send button if you desire to send the message. The system will transmit the message.
    e. Select the Close button. The system will close the Combat Messages dialog box.

7. Prepare and send a SITREP Message.

NOTE: SITREP is a dual-purpose message that generates a situation report message and displays the status of subordinate units. The SITREP displays a table format for ease of entry and a correlated view of received reports. The SITREP table graphically depicts the status of five critical categories that the commander uses to determine the units' status. The SITREP message displays subordinate unit status or provides a detail view of individual platform status two levels down.

    a. Select the Combat Msgs button on the Function Bar or the F3 button on the keyboard.
    b. Select the SITREP tab. The system will display the SITREP template.

NOTE: The combat messages SITREP template displays a color-coded graphical representation of your own situation and the situation of those units/platforms subordinate to you. The color key is as follows: black (not mission capable), red (marginally mission capable with major deficiencies), amber (mission capable with only minor deficiencies), green (full strength), and White (no report received). The SELF-tab displays a color-coded graphical representation of your own situation. The Unit tab displays a color-coded graphical representation of your subordinate platforms situation.

    c. Select the Self-tab if not already selected to EDIT your status. The system will display seven headings and what they represent under each: Self, which is your platform name, Fuel which is the status of your fuel, Mun which is the status of you ammunition, Pers which is status of your personnel, Eqmt which is status of your Equipment, Unit which is status of your unit, and DTG which is the date time group.
    d. Select your Unit/role name which is the name that is under the heading Self. The system will display the Situation Report Details dialog box.

NOTE: The Unit/Role is a dual-purpose button that identifies the Unit/Role associated with the report and displays the Situation Report details. There are three push buttons in this dialog box which are: Edit which displays the SITREP long form message template, the Close button which closes the dialog box, and the Help button which gives help on the Situation Report Details dialog box.

    e. Select Close. The system will close the dialog box and display the Ops (Map) Screen.
    f. Select the Combat Msgs button on the Function Bar or the F3 button on the keyboard. The system will display the Combat Messages dialog box.
    g. Select the SITREP Tab. The system will display the SITREP tab group.
    h. Select the White box with a W in it under the heading Fuel. The box will turn Black with a B in it.

NOTE: Each time you click on that box it will change color. Select the color to indicate what the status is for that category. Each Category is done the same way.

    i. Select the Unit Tab. The system will display the Unit Situation.

NOTE: If you are the Platoon Sergeant or Platoon Leader you will have your subordinate platform's status under your own.

    j. Select the Save button. The system will display Sit Rep saved at and the DTG.
    k. Select the Send button if you desire to send the message. The system will transmit the message.
    l. Select the Close button. The system will close the Combat Messages dialog box and return to the Ops screen.

STP 11-25C13-SM-TG

Performance Measures	GO	NO GO
1. Select the Combat Msgs button.	——	——
2. Prepare and send a SALT message.	——	——
3. Prepare and send a fire mission message.	——	——
4. Prepare and send a MEDEVAC message.	——	——
5. Prepare and send an NBC 1 message.	——	——
6. Prepare and send a Check Fire message.	——	——
7. Prepare and send a SITREP message.	——	——

Evaluation Guidance: Score the Soldier GO if all steps are passed. Score the Soldier NO-GO if any step is failed. If the Soldier scores NO-GO, show him what was done wrong and how to do it correctly.

STP 11-25C13-SM-TG

PERFORM STARTUP PROCEDURES FOR FORCE XXI BATTLE COMMAND BRIGADE AND BELOW (FBCB2) VERSION 3.4
171-147-0002

Conditions: Given a vehicle with an operational Force XXI Battle Command Brigade and Below (FBCB2) system with software version 3.4 loaded, SINCGARS, PLGR, EPLRS (if equipped), and TM 11-7010-326-10.

Standards: Start-up the FBCB2 system and its components for operation in accordance with the TM. As a minimum, you must: Start-up the PLGR, EPLRS if equipped, INC, SINCGARS, Applique+V4 computer, and perform Session Manager functions.

Performance Steps

CAUTION: Apply power to the FBCB2 system in the correct sequence to avoid damage to the equipment. Ensure that the Appliqué+ V4 computer and all interface equipment are off when starting the vehicle.

NOTES:

(1) It is essential that the equipment operator apply power in the correct sequence. By not following an orderly progression, the Appliqué+ V4 computer may not see all the equipment. This will cause a bad start-up and the TIM system will not work correctly.

(2) PLGR display results shown in Table 2-7 in the TM are examples only. Actual results will vary with date, time, and location.

   1. Perform PLGR start-up procedures.

NOTE: PLGR display results are examples only. Actual representation/values will differ.

      a. Press the ON BRT [1] button for two (2) seconds. The PLGR powers up and runs a diagnostics test.

NOTE: It may take as long as 15 minutes for the PLGR to acquire satellites. Ensure the PLGR has a good Line-Of-Sight (LOS) to the sky.

      b. Press the down arrow [5] button. The display will show the Zulu time, Date, TFOM, Speed too slow.
      c. Press the down arrow [5] button three more times or until FOM is displayed.

   2. Perform EPLRS start-up procedures.

NOTE: If the vehicle is not equipped with EPLRS go to step 3.

      a. Turn EPLRS POWER switch on the Receiver -Transmitter, Radio to ON + AUDIBLE.

NOTE: If the green Power indicator is illuminated, proceed to self-test. If the red Alarm indicator is illuminated, follow Unit SOP to load EPLRS COMSEC before proceeding to self-test.

      b. EPLRS performs self-test. On the URO, after 5 seconds, T and URO-OK are displayed in the MODE and MSG descriptor fields.
      c. Press repeatedly the User Readout (URO) RCVD button, until the EAST/BRG and NORTH/RNG fields of the URO display either: If under MSG the following is displayed: @C or @S. This is a good fill. If under MSG the following is displayed: @1, @3, or @4. This indicates improper fill. Reload COMSEC per Unit SOP. If under MSG the following is displayed: @0. This indicates no fill. Reload COMSEC per Unit SOP.
      d. EPLRS and URO, observe whether the OUT-OF-NET red indicator light(s) are lit. If no red light(s) are lit, the EPLRS is correctly initialized. If a steady (or blinking) red light is displayed, refer to the EPLRS TM before proceeding.

STP 11-25C13-SM-TG

Performance Steps

NOTE: It may take several minutes for EPLRS to enter the Net.

    e. Verify that your URO radio Set ID (RSID) is the same as your FBCB2 setting (displayed under Admin/Platform Settings/Misc tab). If correct RSID, proceed with SINCGARS ASIP start-up. If wrong RSID, zeroize the EPLRS and wait 30 seconds. Turn on the EPLRS and enter "_ _" in MSG field. Enter in your FBCB2 RSID and guard channel. Press SEND key. Confirm changes were accepted. Reload COMSEC, then go back to step b.

3. Perform INC start-up procedures.

NOTE: At User read out (URO), after 5 seconds MODE will display T and URO-OK in mode and Msg Descriptor field.

    a. At the Vehicular Amplifier Adaptor (VAA), set the CB1 Power toggle switch to ON (up) position. The DS1 power light is illuminated.

4. Perform SINCGARS ASIP start-up procedures.
    a. At the Receiver/Transmitter (R/T), set the function switch to SQ ON (i.e., Squelch On) and verify settings. WAIT first appears, then the following is displayed: PWR - (LO, M, HI), MODE - (FH), CHAN- (1) per SOP, CMSC - (CT)
    b. Verify that COMSEC Crypto key is loaded. If COMSEC Crypto is loaded, proceed. If COMSEC Crypto is not loaded, follow proper Unit SOP to load Crypto before proceeding. (Refer to SINCGARS TM if necessary)

NOTE: If the COMSEC Crypto key is not loaded, FILL1 will be displayed and a steady tone will be heard over the radio. To enable FBCB2 to communicate, your SINCGARS ASIP must be set to the correct Net ID displayed under the Admin F6 / Platform settings/Misc tab.

    c. Verify that Net ID Frequency displayed is the same as the FBCB2 setting. Press FREQ button on the Receiver/Transmitter (R/T) keypad to display SINCGARS Net ID frequency. If correct frequency is displayed, go to next step d. If wrong frequency is displayed, set the FBCB2 Data Net Frequency into radio before proceeding. (Refer to SINCGARS TM if necessary.)
    d. Verify PCKT mode. Press the 4 button on the R/T keypad, then press the 7 button repeatedly until PCKT is displayed. PCKT Data Mode is set and verified when PCKT is displayed.

5. Perform Computer start-up/Log in on the AN/UYK-128(V).

NOTES:

(1) This step describes the start-up/Log in procedures for the AN/UYK-128(V) computer. The PLGR, EPLRS (if equipped) INC, and SINCGARS ASIP (if equipped) must be fully operational before powering-up the AN/UYK-128(V) computer.

(2) Ensure KU is connected to the AN/UYK-128(V) computer prior to start-up/log in.

    a. Ensure the vehicle primary power is on. Set circuit breaker / switch on PU to on. The circuit breaker/ switch is pointed toward the centerline of the PU when set to the ON position.
    b. Press the Display Unit (DU) PWR button for up to (4) seconds and release after the PWR green Light Emitting Diode (LED) illuminates. PWR, D ISP, and CPU green light LEDs illuminate. Start-up continues automatically until DU displays the Session Manager Screen with Task bar on the bottom.
    c. Select Cancel Time out button on the Ops Auto-Login dialog box. The Ops Auto-Login dialog box closes after the Cancel Time out button is selected. The Comm initialization dialog box is displayed on the Session Manager screen and continuously updates during Comm start-up. The Router initialization dialog box is displayed and will not update until Ops button is selected.

NOTE: The Count Down Timer has a 20 second time limit. If the timer is allowed to go to zero, the FBCB2 process to go online will begin automatically.

STP 11-25C13-SM-TG

Performance Steps

d. Check the color of both GPS and Router dialog boxes (when Comm and GPS cycle is complete) to determine the operational status. Green color-coding indicates that the equipment is fully mission capable. Yellow color-coding indicates degraded operation. Red color-coding in the GPS and/ or Router dialog box indicates that initialization has failed and that the GPS and/or Router is not responding.

NOTE: Red or Yellow color-coding indicates a problem to be addressed before proceeding.

e. Select the Done button on both the GPS and Router dialog boxes when initialization is complete. The GPS and Router dialog boxes close.

NOTE: Always verify that the correct Unit/Role ID is displayed before proceeding with Login. Otherwise, the user will not be able to receive his incoming FBCB2 messages.

f. Check the Unit/Role in the Function Bar located in the lower right hand corner of the Session Manager screen. If correct Unit/Role is displayed, proceed to next step. If the wrong Unit/Role is displayed, perform configure role setup (refer to the user's manual) before proceeding.
g. Select the Start button. The Start option menu is displayed.
h. Select the Login option. The Ops Login dialog box is displayed.
i. Type in the password. Asterisks are displayed as the password is typed followed by a blinking cursor in the Password text box.
j. Select the Continue button. The Ops Login dialog box closes.

NOTE: Perform clear logs and queues as per SOP. Perform touchscreen calibration if required.

k. Select the Ops button in the Session Manager Function Bar. The system goes online. The Ops Main Screen with the FBCB2 Display Process dialog box is displayed.

6. Perform Touchscreen Calibration on the AN/UYK-128(V) computer.

NOTE: The Calibrate Touchscreen option is used to realign Display Unit touchscreen sensors with the FBCB2 software. This procedure is performed after the AN/UYK-128(V) start-up/Login is completed.

a. Select the Start button. The Start option menu is displayed.
b. Select the Settings option. The Settings option menu is displayed.
c. Select the Touchscreen option. The calibration touchscreen is displayed with a target bull's-eye at the lower left hand corner.
d. Select the center of the bull's-eye with the stylus. The calibration touchscreen is displayed with a target bull's-eye at the upper right hand corner.
e. Select the center of the target bull's-eye with the stylus. The calibration touchscreen is displayed with a target bull's-eye at the lower right hand corner.
f. Select the center of the target bull's-eye with the stylus. The calibration screen closes. The osc_touch_calibrate.ksh dialog box is displayed.
g. Type the letter "y." The letter y is displayed at the ENTER prompt.
h. Select an enter button or key. The osc_touch_calibrate.ksh dialog box closes.

7. Perform Clear Logs and Queues on the AN/UYK-128(V) Computer.

NOTE: The Processor Unit will operate more efficiently (less chance of a software slowdown or need to reboot) if logs and queues are cleared regularly. This procedure is performed after the AN/UYK-128(V) start-up/Login is completed. Performed while off line but after login.

a. Select the Start button. The Start button option is displayed.
b. Select the FBCB2 option. The FBCB2 option menu is displayed.
c. Select the Clear Logs and Queues option. The Clear Logs and Queues dialog box is displayed.
d. Under Select Items to Clear, select option(s) by selecting button next to corresponding option. Option(s) button selected is shown with check mark.

STP 11-25C13-SM-TG

Performance Steps
    e. Select the Apply button to clear selected option(s). Clear Logs and Queues Status dialog box opens.
    f. Select Close button when message COMPLETED CLEAR LOGS AND QUEUES OPERATION is displayed. The Clear Logs and Queues Status dialog box closes.
    g. Select Close button in Clear Logs and Queues dialog box. Clear Logs and Queues dialog box closes.

Performance Measures   <u>GO</u>   <u>NO GO</u>

1. Perform PLGR start-up procedures.   ——   ——

2. Perform EPLRS start-up procedures.   ——   ——

3. Perform INC Start-up procedures.   ——   ——

4. Perform SINCGARS ASIP start-up procedures.   ——   ——

5. Perform Computer Start-up/Log in on the AN/UYK-128(V).   ——   ——

6. Perform Touchscreen Calibration on the AN/UYK-128(V) computer.   ——   ——

7. Perform Clear Logs and Queues on the AN/UYK-128(V) Computer.   ——   ——

Evaluation Guidance: Score the Soldier GO if all steps are passed. Score the Soldier NO-GO if any step is failed. If the Soldier scores NO-GO, show him what was done wrong and how to do it correctly.

STP 11-25C13-SM-TG

APPLY MESSAGE ADDRESSING FEATURES IN FBCB2 VERSION 3.4
171-147-0005

Conditions: Given a vehicle with an operational Force XXI Battle Command Brigade and Below (FBCB2) system with all of it's components with software version 3.4 loaded and a completed message.

Standards: As a minimum: You must apply the Message Addressing settings so that the message can be transmitted to other platforms.

Performance Steps

NOTE: Message Addressing allow the operator to set the default message options for each message subtype. These default message options should be standardized and included in the unit's TACSOP. This will help speed the message sending process. If the settings and addresses for each message subtype are set ahead of time, that is one less thing that the operator to do prior to sending the message. To take advantage of this capability, the default settings should be set for each type of message that the operator is going to use, especially for Combat Messages. Even if the Default Message addressing has been set for all the message types, the message address can be changes from within the Message itself by selecting the Message Addressing button. Operators should not change the default settings unless the unit TACSOP directs it.

1. Set the Default Message Addressing settings.
    a. Select the Messages button from the Function bar or the F4 button on the keyboard. The Messages dialog box will display.
    b. Select the Create tab.

NOTE: The Create tab group contains the following options: (1) Message Type Radio buttons, which include Orders/Requests, Fires/Alerts, Reports, and Overlays. (2) Message Sub Type panel, which shows a list of the message subtypes. (3) Set Default Message Addressing button for setting the default message options for message sub types. (4) Edit Address Groups button for creating and editing their own address groups. (5) Execute, Close, and Help buttons

    c. Select a Message Type Radio button (Orders/Requests, Fires/Alerts, Reports, Overlays). The radio button will highlight for the message type that was selected and a list of message subtypes will appear in a panel to the right.
    d. Select a message subtype from the list by left clicking on the subtype. The subtype will highlight.

NOTE: If there is a scroll bar on the right of the panel, then there are more subtypes below.

    e. Select the Default Message Addressing button. The Message Addressing dialog box will display.

NOTE: This Message Addressing dialog box contains two tabs: The Message Settings tab and the Addresses tab.

    f. Select the Message Settings tab.

STP 11-25C13-SM-TG

Performance Steps

NOTES:

(1) This tab contains four areas: (1) Precedence for Emergency Cmd, Flash, Immediate, Priority, and Routine. (2) Acknowledge- Check boxes for "MA" (Machine Acknowledge), "OA" (Operator Acknowledge) and "OR" (Operator Response). "MA," "OA," and "OR" may be grayed out depending on the message subtype that was selected. (3) Security Level- Sets the Security level of an outgoing message. (Grayed out. Only available when FBCB2 System is in the Classified mode). (4) Perishability DTG - Indicates when the message is no longer valid and is not a function of the "Default Message Addressing settings.

(2) The "Precedence" radio button allows the user to set the message precedence for the selected message subtype. The system places the selected precedence in the message header of the outgoing messages. This directs the receiving FBCB2 System to place the message into the appropriate incoming message queue. The message precedence also ties to the priority the INC uses to forward the message for delivery.

    g. Select Precedence radio button. The radio button will highlight.
    h. Select an Acknowledge check box. The checkmark will appear in the box.

NOTES:

(1) Use the Acknowledge function if the user wants the receiving station to confirm receipt of the message, or if you want the operator to respond to the message. The Acknowledge function also activates the incoming message queue of the receiving station.

(2) "MA" - Machine Acknowledge - The receiving system transmits an automatic response to the sending system when the message is received. "OA" - Operator Acknowledge - The receiving system transmits an automatic response to the sending system when the operator opens the message. There is an audible alert associated with this type of response as long as it is not muted. "OR" - Operator response - The operator is required to indicate compliance and give a short written reply. There is an audible alert associated with this type of response.

    i. Select the Security level radio button associated with the message. The radio button will highlight. (If in Secret mode).

NOTE: The "Security Level" allows the user to select the classification of an outgoing message. This only works on a system that is configured as Secret. An unclassified user cannot send a secret message.

    j. Select the Addresses tab. The Addresses tab options will display.

NOTE: This tab contains the following options: (Add) - adds addresses, (Delete) - deletes selected addresses, (Delete All) - deletes all selected addresses, (Search) - Executes a search for entered text, (Keypad) - Access a victual keyboard, (OK) - Accepts changes and closes the dialog box, (Apply) - Applies changes and keeps the dialog box open, (Restore Defaults) - Restores message addressing settings back to original defaults, (Close) - Closes the dialog box without saving changes, (Help) - Accesses context sensitive Help.

    k. Select the Select From: pull-down arrow. The system will display the Select From menu.

NOTE: This menu contains three options: (1) Master Address Book- contains the addresses of all units in the current UTO, the user cannot add or delete addresses from the Master address Book, (2) User Address Groups - Groups created by the user, Battalion Address Book - Contains the addresses within the units platform.

    l. Select the Master Address Book. The Master Address Book will Highlight and display in the Select From: text box and all the addresses in the current UTO will appear in the pane below.

16 February 2005

3 - 123

STP 11-25C13-SM-TG

Performance Steps

NOTE: The user can select an address by selecting it from the list or by typing the name in the search text box.

    m. Select a unit from the pane by left clicking on the + sign to expand the major unit and left clicking on the unit, or type the name in the Search text box and select the Search button. The name will highlight in the Select From: pane.
    n. Select the Add button. The name will appear in the Address pane on the right.

NOTE: Do the same for each address that needs to be added.

    o. Select the Add Addresses pull-down arrow under the Addresses tab name. The system will display Action Addresses and Info Addresses.

NOTE: Action addresses are addresses that need to respond in some way. Info Addresses are addresses that are getting the message as information and do not include an acknowledgement. The default is Action Addresses.

    p. Select either Action or Info addresses. The option will highlight and appear in the Add Addresses text box.

NOTE: An address cannot exist in Action Addresses and Information Addresses simultaneously.

    q. Select the Apply button. The system will apply the entries and keep the dialog box open.
    r. Select the close button. The system will return to the Messages dialog box.

2. Create Address Groups.

NOTE: Each platform has several predefined user address groups. The platform's role determines which groups exist, and the addresses in each group. The address groups consist of groups that the platform belongs to by doctrine, groups that the platform's echelon owns, and groups that that the platform's parent organization owns. This does not prevent the operator from generating additional groups to fit specific needs. The operator has the ability to add or delete an address, or delete an entire group.

    a. Select the Create tab.
    b. Select the Edit Address Groups button. The Edit Address Groups dialog box displays.

NOTE: The following functions are: (New Group) - Adds a new group to the right hand window; (Rename Group) - Renames the selected group in the right hand window to the new name; (Delete Group) - Deletes the selected group from the right hand window; (Add Address) - Adds the address from the left hand window to the selected group in the right hand window; (Delete Address) - Deletes the selected address from the right hand window; (Search) - Executes a search for entered text. Data can be entered using either the virtual keyboard or the computer keyboard. (Name Text Box) - Allows the user to manually type text. The left window displays a listing of all addresses. The right window displays the folder list. If there are no user address groups, there are no folders listed.

    c. Select the Name text box by placing the cursor in the text box and left clicking and typing the name that the user wants to call the folder, or Select the Virtual Keyboard and type the name using the mouse pointer and select OK. The name will appear in the text box and the New Group button will become active.
    d. Select the New Group button. The folder will be created in the window on the right side.
    e. Select the Folder that was just created by highlighting it. The Delete Group button will become active.
    f. Select the Search text box and type the name of the unit that the user wants to add to the new group, or select the Virtual Keyboard and type the name and select the OK button. The Name will highlight on the left window and the Add Addresses button will become active.

NOTE: If the Add Addresses button does not become active, click on another name then back on the correct name in the left window and the button should become active.

STP 11-25C13-SM-TG

Performance Steps

    g. Select the Add Address button. The address will be added to the Folder on the right side window.

NOTE: Continue these same steps until all of the addresses are added.

    h. Select the Folder that was created with all the addresses in it.

NOTE: If there is a plus sign next to the folder, click on it to open the folder to view all of the addresses within the folder.

    i. Select a name that is in the folder by highlighting it. The Delete Address button becomes active.

    j. Select the Delete Address button. The address is deleted from the window on the right.

NOTE: Continue this until all the addresses that the user wants to delete are deleted. If the user wants to delete a whole group, it is done the same way. Highlight the folder name and the Delete Group button becomes active. The user just selects the Delete Group button to delete the whole group.

    k. Select the OK button. The system will save and exit the Edit Address Groups dialog box and return to the Messages dialog box.

    l. Select the Close button. The system will return to the Ops screen.

Performance Measures	**GO**	**NO GO**
1. Set the Default Message Addressing settings.	——	——
2. Create Address Groups.	——	——

Evaluation Guidance: Score the Soldier GO if all steps are passed. Score the Soldier NO-GO if any step is failed. If the Soldier scores NO-GO, show him what was done wrong and how to do it correctly.

STP 11-25C13-SM-TG

PERFORM MESSAGE MANAGEMENT USING FBCB2 VERSION 3.4
171-147-0006

Conditions: Given a vehicle with an operational Force XXI Battle Command Brigade and Below (FBCB2) system with software version 3.4 loaded. To better organize FBCB2 information you have received, you need to manage your messages.

Standards: As a minimum: you must rename, delete files and folders, and move files from one folder to another by using the Manage tab on the Messages function.

Performance Steps

1. Select the Messages button from the Function Bar or the F4 button on the keyboard.

2. Select the Manage Tab. The system will display the Manage fields and push buttons.

NOTE: The Manage tab allows the user to create, delete, move, and rename folders and delete messages. There are eight push buttons consisting of: (New Folder), which creates a folder; (Delete), which deletes folders and files; (Rename), which rename folders and files; (Move), which moves files from one folder to another; (Execute), which activates a message; (Grayed Out); (Close), which closes the Messages dialog box; and (Help), which gives help on the Manage Tab; and a Virtual Keyboard.

   a. Ensure that all of the message type boxes are checked under the Msg Type so that all the folders and files will be displayed in the folder window on the right.
   b. Select the + sign next to a folder that the user wants to manage or double-click the folder name, which will also open the folder. The folder will open and display all the files within that folder.

3. Rename a file.
   a. Select one of the files listed under the folder that needs to be renamed. The file will be highlighted.
   b. Select the Name: text box at the bottom of the Messages dialog box. The cursor will be blinking inside the text box
   c. Type the new name. The name will display in the text box.
   d. Select the Rename button. The file name will be changed and displayed under the folder in Alphabetical / Numerical order by file names.

NOTE: If the user typed a name that is not valid, a Rename Error dialog box saying, "This name is not valid, please re-enter the new message name!" with an OK button will appear.

4. Rename a folder.
   a. Select the Folder name from the folders window. The folder name will be highlighted.
   b. Select the Name: text box at the bottom of the Messages dialog box. The cursor will be blinking inside the text box.
   c. Type the new name for the folder. The name will display in the text box.
   d. Select the Rename button. The Folder name will change and be displayed in Alphabetical/ Numerical order by Folders.

5. Delete a file or folder.

NOTE: The steps for deleting a file or folder are the same.

   a. Select a file from one of the folders. The file will be highlighted.
   b. Select the Delete button. The Delete File dialog box will be displayed saying "You are about to delete message "name of File or Folder" and its associations, continue with this action?" with a OK and Cancel button.
   c. Select the OK button. The file or Folder that the user selected will be deleted.

STP 11-25C13-SM-TG

Performance Steps

6. Move a file from one folder to another folder.
   a. Select a file from one folder. The file will be highlighted.
   b. Select the Move button. The Choose Destination Folder dialog box will display with all the folders that have been saved.
   c. Select the folder that the user wants to move the file to. The Folder will be highlighted.
   d. Select the OK button. The file will be displayed in the new folder.
   e. Select the Close button. The system will return to the Ops screen.

Performance Measures	GO	NO GO
1. Select the Messages button from the Function Bar or the F4 button on the keyboard.	——	——
2. Select the Manage Tab. The system will display the Manage fields and push buttons.	——	——
3. Rename a file.	——	——
4. Rename a folder.	——	——
5. Delete a file or folder.	——	——
6. Move a file from one folder to another folder.	——	——

Evaluation Guidance: Score the Soldier GO if all steps are passed. Score the Soldier NO-GO if any step is failed. If the Soldier scores NO-GO, show him what was done wrong and how to do it correctly.

STP 11-25C13-SM-TG

PREPARE/SEND OVERLAYS USING FBCB2 VERSION 3.4
171-147-0007

Conditions: Given a vehicle with an operational Force XXI Battle Command Brigade and Below (FBCB2) system with software version 3.4 loaded, an operation order (OPORD), and the commander's guidance. Default Message Addressing settings have been set.

Standards: As a minimum, you must: create an overlay or select an overlay, change or edit an overlay, save the overlay, and send an overlay.

Performance Steps

NOTE: Operational plans, operational concepts, units and weapon symbols, objectives, boundaries, routes of march, and other control measures can be created in a digital overlay to provide command and control. Digital Overlays can be created faster with greater degree of accuracy and can be edited and quickly disseminated to other members in the BDE. Some overlays automatically post to the map SA area and generate an audible alert when they are received which is designed to prompt the operator to look at their display for life threatening obstacles. The SA display on the FBCB2 system can be tailored to show up to four distinct layers of information. Any combination of these layers can appear on the SA display simultaneously. The Four electronic layers consist of the currently displayed overlay, previously loaded overlay, any SA elements that the system has been set to display, and the map which forms the base of the display. Layer 1 is the Map Display. Layer 2 is the SA display. Layer 3 is the Loaded Overlays. Layer 4 is the current overlay.

1. Select the Messages button from the Function Bar or the F4 button on the keyboard. The system will display the Message dialog box.
    a. Select the Create tab. The system will display the Create Tab options.
    b. Select the Overlays radio button under the Msg Type field. The Overlays radio button will highlight and show a list of available Overlay types.

NOTE: If there is a scroll bar to the right of the list, then there are more types of overlays to view.

   c. Select an Overlay type from the list by highlighting it.

NOTE: If the user doesn't set the Default Message Addressing Settings at this time, then the user can set them before the message is sent.

   d. Select the Execute button. The system will display the Overlay Toolbox dialog box.

NOTE: The system will display Three tabs: the Overlay, Object, and Group Setup tab. The Overlay tab is used to select the Overlay type, save an overlay, and reference specific information about the overlay. The Object tab is used to add and edit symbology to the map SA area. The Group Setup tab allows the operator to create a customized group of icons for specific missions; for example, Operations symbology or Fire Coordination symbology.

2. Select the Overlay tab if it is not already selected. The system will display the Overlay tab fields.

NOTE: The Overlay tab allows the user to set the basic information for the overlay. This is where the overlay type, Operations Order or Operations Plan reference, storage data, version, message size, statistics and additional comments are set for the overlay. There are eight push buttons at the bottom of the Overlay tab: Save button, which saves the overlay; Delete, which deletes the overlay; Message Addressing, which is used to set the address where the overlay will be sent; Save As, which is used to save the overlay initially; Print (unavailable); Send, which is used to send the overlay; Keep overlay displayed, which is used to keep the overlay displayed on the map when you close the Overlay toolbox; Close, which is used to close the Toolbox; and Help for help on the overlay toolbox.

   a. Select the Overlay Type pull-down arrow to display the different overlay types. The system will display the Overlay type list.

Performance Steps

NOTE: The user may have to scroll up to view the Overlay Type pull down arrow. The Overlay type that was selected in step 4 should be displayed in the Overlay Type text box. If not, then select it now.

    b. Use the Scroll bar on the right side of the Overlay Toolbox and scroll down to the Operation field.

    c. Select the Operation Set radio button. The system will display the DTG text box, DTG button, and Identification text box.

NOTE: The Operation option button is used to enter/change the Operation DTG and Identification number.

    d. Place the cursor in the DTG text box and type the date time group in with the keyboard or select the DTG button and use the + or - keys to enter the DTG or select the NOW button for the current DTG and select the OK button. The system will display the DTG in the DTG text box.

    e. Place the cursor in the Identification Num text box and enter the Identification number (numeric 00 to 99) in the text box. The system will display the number in the Identification Num: text box.

NOTE: This number ties the overlay to an Operations Order or an Operations Plan.

    f. Scroll down and select the Recalc Size button to calculate the size in bytes and display the overlay file size in the Message Size information box.

NOTE: This function only works after you have placed symbols on the overlay. This is to inform the operator of how large the overlay size is in Bytes. See the equipment operator's manual or pocket guide for maximum overlay size that can be transmitted.

    g. Scroll down and select the Details button under the Statistics field. The system will display the Objects Details dialog box.

NOTE: There will not be data in the Object Details dialog box until the operator has placed objects on the overlay. This function is used to quickly center on an object and view it's attributes.

    h. Select the Close button to close the Objects Details dialog box. The system will go back to the Overlay Toolbox.

    i. Select the Comment Set radio button to activate the comments field. The system will display the Comment text box and make the Virtual Keyboard button become active.

NOTE: The comments field is limited to a maximum of 200 alphanumeric characters that is used to make comments about the overlay. To add comments, just left click in the text box and type the comment.

    j. Select the Save As button. The system will display the Save As dialog box.

    k. Highlight the folder that the overlay will be saved to, or create a folder with the New folder button.

    l. Type the name of the overlay in the File text box at the bottom of the Save AS dialog box.

    m. Select the OK button. The system will display an Information dialog box stating "The message was saved" with an OK button.

    n. Select the OK button. The system will return to the Overlay Toolbox.

3. Select the Object tab. The system will display the Object tab group consisting of : Group, 2525B, and UTO tabs.

NOTE: The Object tab selects an object, identifies its attributes, (friend, foe, or unknown), and adds it to an overlay. The Object tab displays all the object information and is used to add graphical objects, (create) or revise (edit) a graphical objects attributes, textual labels, or location on the current overlay.

4. Select the Group Sub tab (Default). The system will display the Group sub tab group.

Performance Steps

NOTE: The Group Sub tab displays graphic symbol files and sub-files and is used to select and add icons to the Map SA area. The Graphic symbol files are: Areas, Equipment, Equipment-Air, Fire Support, Lines, Obstacles, Points, Units, and any groups that the user created.

    a. Select the Group pull down arrow. The system will display the graphic symbol types.
    b. Select a graphic type from the list by highlighting it with the mouse pointer. The graphic type will display in the Group text box and a list of graphic sub types will display in the preview windowpane below.
    c. Select a graphic sub type from the list. The selection will highlight and the Add button will become active.
    d. Select the Add button. The system will display the Map with the Create Object function bar to the right and the cursor will become a cross hair.

NOTE: Some objects require more than one point on the map to be inserted. If more than one point is required and the user selects the OK button without selecting the required amount of points, the system will display a Warning dialog box saying " Requires at least Three points." When adding objects that require more then one point, a solid white circle will represent each point.

    e. Select a location on the map where the object will be inserted. The object will appear on the map.
    f. Select the OK button. The Overlay Toolbox will be displayed.

NOTE: Continue steps (1) through (6) for all the graphics that need to be inserted.

    g. Select the Grab button. The Overlay Toolbox dialog box will disappear and the cursor will be replaced with the pencil pointer.
    h. Place the pointer on the map to the top left of the object that the user wants to grab and left click the mouse.
    i. Move the pointer to the bottom right of the object and left click the mouse. A box will appear over the object and the Overlay Toolbox will reappear.

NOTE: The user can select more than one object by dragging the box over several objects.

    j. Select the Edit button. The system will display the Overlay Edit dialog box with three sub tabs containing: Attributes, Labels, and Location.

NOTE: This function is used to modify a graphical object's characteristics, including affiliation, status, color, mobility, size, and designation. Some of these areas will be grayed out depending on the selected object. There is a preview window to view the selected object. The two arrows pointing to the left and right are the scroll arrows to move through the selected objects that were grabbed earlier. The number on the left of the / is the object that is selected for edit and the number on the right of the / is the amount of objects that were grabbed.

    5. Select the Attributes tab. The system will display the Attributes sub tab group.

NOTE: Some functions may be grayed out depending on the object that is selected for edit.

    a. Select the Status pull down arrow. The system will display the Status menu.

NOTE: This refers to whether an object currently exists at the location (Present) or in the future could reside at the location (Planned).

    b. Select the status from the list. The selection will appear in the Status text box.
    c. Select the color pull down arrow. The system will display the colors that can be selected.
    d. Select the Affiliation pull down arrow. The system will display the Affiliation menu.

NOTE: The Affiliation field allows the user to set the affiliation that is associated with the selected object.

    e. Select an affiliation from the menu. The affiliation will display in the Affiliation text box.

STP 11-25C13-SM-TG

Performance Steps

   f. Select the Mobility pull down arrow. The system will display the Mobility menu.

NOTE: The Mobility field allows the user to set what type of mobility (if any) the object is.

   g. Select the Size pull down arrow. The system will display the Size menu.

NOTE: The Size field allows the user to set the size of the object.

   h. Select a size from the menu. The Size will display in the Size text box.
   i. Place a Checkmark in the Check box next to HQ, Task Force, Feint, or Installation depending on what the Icon represents. The user can place checkmarks in more than one check box.
   j. Select the Apply button. The changes will apply and return to the Overlay Edit dialog box.

6. Select the Labels tab. The system will display the Labels tab group.

NOTE: The Labels function allows the user to modify an object's icon labels.

   a. Select the First DTG button. The system will display the DTG keypad.

NOTE: The DTG button allows the user to enter the Date, Time group associated with the object.

   b. Select the Year text box by selecting it with the mouse pointer.
   c. Select the OK button. The system will return to the Overlay Edit dialog box.
   d. Select the + or - key to increase or decrease the year or select the NOW button.

NOTE: Do the same for the Month, Day , Hour, and Minute for each object that is entered on the overlay.

   e. Select the Additional Information text box and type the information or select the virtual keyboard by entering the information with the mouse pointer.

NOTE: The Additional Information function allows the user to enter supplementary facts about the object.

   f. Select the Unique Designation text box and type the specification, or select the virtual keyboard and enter the specification with the mouse pointer and select OK.
   g. Select the Apply button. The system will apply the entries and keep the Overlay Edit dialog box open.

7. Select the Location tab. The system will display the Location Tab group.

NOTE: The Location Tab displays the Current Grid Points of a selected object, whether it is a single point object or a multi-point object. The Location tab facilitates moving single point objects, such as a unit and modifying the points of multi-point objects such as routes. The user can move an object to another location on the map by using the Fill Location button and modify the dimensions of a multi-point object.

   a. Select the right arrow at the bottom left of the Overlay Edit dialog box. The system will cycle to the next object and display the grid coordinates of the object.

NOTE: Each time the user selects the right arrow, it will cycle to the next object.

   b. Select a grid from the list by highlighting it with the mouse pointer. The grid will highlight.
   c. Select the Fill Location button. The Overlay Edit dialog box will disappear and the pointer will become a cross hair.
   d. Select a spot on the map where the user wants to move the object for a single points or a dimension of a multi-point object. The Overlay Edit dialog box will reappear.
   e. Select the Apply button. The object or dimension will move to the new location.

NOTE: Do the same for all the objects or dimensions that need to be moved or modified.

   f. Select the OK button. The system will return to the Overlay toolbox.

8. Select the 2525B. The system will display the 2525B tab group.

16 February 2005

3 - 131

STP 11-25C13-SM-TG

Performance Steps

NOTE: The 2525B sub-tab function allows the user to acquire Warfighting symbols, Tactical graphics, weather and signal intelligence symbols. The 2525B sub tab works in the same way as the Group sub tab. Choose from the Dimension, Type, and Sub-type menus in order to select the desired symbol.

    a. Select the Dimension pull down arrow. The system will display the Dimension menu.
    b. Select a symbol from the list. The symbol name will display in the Dimension text box.
    c. Select the Type pull down arrow. The system will display the Type menu.
    d. Select a type from the menu. The Type name will display in the Type text box.
    e. Select the Subtype pull down arrow. The Subtype menu will display.

NOTE: The Subtype menu may or may not be available depending on the Dimension or type that was selected in steps 2 and 4.

    f. Select a Subtype from the list. The name will appear in the Subtype menu.
    g. Select the Attributes button. The Symbol Attributes dialog box will display.

NOTE: The Attributes button is done exactly like the Attributes Tab in step a, sub step 11 through 20.

    h. Select the OK button after completing the Attributes. The Overlay Toolbox will reappear.

NOTE: Do the same for all the symbols that are placed on the overlay.

  9. Select the UTO sub-tab. The UTO sub tab group will display.

NOTE: The UTO sub tab allows the user to select unit symbols from various Unit Task Organizations (UTO) for inclusion in the overlay.

    a. Select the Search button. The system will display the UTO Search dialog box.
    b. Type the Name of the unit in the search text box with the keyboard or the virtual keyboard. The name will display in the search text box.
    c. Select the Search button. The unit will be highlighted.
    d. Select an Organization in the Unit column. The Organization will be highlighted and the unit symbol appears in the Preview pane in the bottom left hand corner.
    e. Select the OK button. The system will return to the UTO sub tab with the unit name under the Unit field and the symbol in the preview pane.
    f. Select the Attributes button and enter the attributes for the symbol and select Octet system will return to the Overlay Toolbox.
    g. Select the Add button. The system will display the unit symbol on the overlay.
    h. Select the OK button. The system will return to the Overlay Toolbox.

NOTE: Continue sub steps (1) through (8) until all symbols are on the overlay.

  10. Select the Overlay tab. The system will display the overlay tab fields.

NOTE: Scroll up to the Overlay Type. Ensure that it still shows the Overlay type that was selected at the beginning. Scroll down to the Storage field. Ensure that the Folder and File display the correct name. Scroll down to Version to view the Creation DTG, Revision DTG, and Revision. Scroll down to the Recalc Size button under the Message Size field. Select the Recalc button to display the size of the overlay in bytes. The user can center on an object by selecting the Details button, then highlighting the object from the list, and selecting the Center on Object button. The user can set the Message Addressing settings and re save the overlay from here.

  11. Select the Group Setup Tab. The system will display the Group Setup tab fields and buttons.

NOTE: The Group Setup Tab creates user-defined object group files and sub-files. User-defined object groups can be developed and used to expedite the creation of overlays.

    a. Select the Group down arrow. The system will display the Group menu including the user-defined groups if there were any created.

STP 11-25C13-SM-TG

Performance Steps
      b. Select the Group Name text box. The text box will highlight.
      c. Type a name for the new group folder. The name will appear in the Group Name text box.
      d. Select the New Group button. The system will display the newly created folder in the Group text box.

12. Select the Object tab then the Group sub tab.
    a. Select the Group down arrow. The system will display the Group Menu including the group folder that was just created.
    b. Select a group folder from the list. The sub file names will display in the Group window.
    c. Select a group sub file from the list of sub files. The sub file will be highlighted and the icon will be displayed in the preview window.

13. Select the Group Setup tab. The newly created folder should still be in the Group text box.

NOTE: The Add Icon button will be active now.

    a. Select the Add Icon button. The sub file will be displayed under the New folder.

NOTE: Repeat this process until all desired icons are in the new Group folder. After saving Object Groups, they remain in the system until the user deletes them from within the Group Setup tab or by clearing Logs and Queues.

    b. Select the new folder that was created by highlighting it in the Group text box.
    c. Select the Save Groups button. The system saves the Group.

NOTE: If the user wants to delete a group sub file, highlight the sub file and select the Delete Icon button. If the user wants to delete a group, highlight the Group name and select the Delete Group button.

    d. Select the Close button. The system returns to the Ops screen.

Performance Measures	**GO**	**NO GO**
1. Select the Messages button from the Function Bar or the F4 button on the keyboard.	——	——
2. Select the Overlay tab if it is not already selected.	——	——
3. Select the Object tab.	——	——
4. Select the Group Sub tab (default).	——	——
5. Select the Attributes tab.	——	——
6. Select the Labels tab.	——	——
7. Select the Location tab.	——	——
8. Select the 2525B.	——	——
9. Select the UTO sub-tab.	——	——
10. Select the Overlay tab.	——	——
11. Select the Group Setup Tab.	——	——
12. Select the Object tab then the Group sub tab.	——	——
13. Select the Group Setup tab.	——	——

Evaluation Guidance: Score the Soldier GO if all steps are passed. Score the Soldier NO-GO if any step is failed. If the Soldier scores NO-GO, show him what was done wrong and how to do it correctly.

STP 11-25C13-SM-TG

PREPARE/SEND REPORTS USING FBCB2 VERSION 3.4
171-147-0008

Conditions: Given an operational Force XXI Battle Command Brigade and Below (FBCB2) system with software version 3.4 loaded. You have received an operation order (OPORD) or fragmentary order (FRAGO) to conduct a tactical operation. Default Message Addressing settings have been set.

Standards: As a minimum: you must prepare and send one each of the following reports: Airborne Artillery, Personnel Status Report and a Freetext message. All mandatory fields must be completed.

Performance Steps

NOTE: The Airborne Artillery FCR, Bridge Report, Chemical Downwind Report, Contact Report, Effective Downwind Message, Engagement Report, Initial Airborne Artillery FCR, Land Minefield Laying Report, Medical SITREP, Mortuary Affairs, NBC3, NBC4, Obstacle Report, Position Report, and Supply Point Status Report are all done the same as the Airborne Artillery FCR. For LOG Report, refer to task 171-147-0004 (Prepare/Send a Platoon Logistical Status Reports Using FBCB2 Version 3.4) and task 171-147-0015 (Prepare/Send a Vehicle Logistical Status Reports Using FBCB2 Version 3.4). The Land Route is done the same as task 171-147-0023 (Employ NAV Functions Using FBCB2 Version 3.4). NBC1 is done the same as task 171-147-0001step 5(Prepare/Send Combat Messages Using FBCB2 Version 3.4). The Personnel Status Report is done the same as task 171-147-0022 sub- step f. (Employ Apps Functions Using FBCB2 Version 3.4). The Situation Report is done the same as task 171-147-0001, step 7 (Prepare/Send Combat Messages Using FBCB2 Version 3.4). The Spot Report is done the same as task 171-147-0001, step 2 (Prepare/Send Combat Messages Using FBCB2 Version 3.4).

1. Prepare and send an Airborne Artillery FCR Report.
    a. Select the Messages button on the Function Bar or the (F4) button on the keyboard. The system will display the Message dialog box.
    b. Select the Create tab.

NOTE: Default Message Addressing Settings for all the Reports should be set prior to the PCC/PCIs. Refer to Task 171-147-0005 Apply Message Addressing Features in FBCB2 Version 3.4.

    c. Select the Reports radio button form the Msg Type Field. The Reports radio button will highlight.

NOTE: The radio buttons are the diamond shaped boxes under the Msg Type field.

    d. Select the Airborne Artillery FCR option from the Message windowpane. The Airborne Artillery FCR message will highlight.
    e. Select the Execute button. The system will display the Create: Airborne Artillery FCR Template
    f. Select the Outline Tab.

NOTE: All of the fields that are red in color and have an "M" in front of them are mandatory fields. These fields must be completed in order for the message to be sent. If the user tries to save or send the message without entering data into the mandatory fields, they will get the error message: "The mandatory field Scan URN is not complete" with an OK button. All the fields in Black are optional. Some Black fields may be mandatory per SOP.

    g. Select the "M" Mission Selection field from the Outline windowpane by highlighting the field. The Mission Selection in the right windowpane will also highlight.

NOTE: The Outline windowpane is a quick way of getting to the option that the user wants to enter data.

    h. Select the Mission Selection pull-down arrow in the right windowpane. The options will be displayed.
    i. Select an option from the list. The option will be displayed in the text box.

Performance Steps
- j. Select the mandatory field Scan URN from the Outline tab. The option will highlight in the Outline windowpane as well as the right windowpane.
- k. Select the pull-down arrow. The system will display the Unit Name dialog box.
- l. Select a Unit Name from the list or type it in the search text box to find the Unit, then select it from the list. The Unit name will be displayed in the text box.
- m. Select the mandatory field Scan Time from the Outline windowpane.
- n. Select the Scan Time DTG button. The DTG Keypad dialog box will be displayed.
- o. Select the + or - buttons to increase or decrease the Day, Hour, and Minute or select the NOW button and OK button.
- p. Select the mandatory field Location from the Outline windowpane.
- q. Select the Location pull-down arrow. The system will display the Location menu.

NOTE: This function allows the user to enter the location using the Keyboard (Kbd), Laser Range Finder (LRF), Map by using the mouse pointer (Map), Virtual Keyboard (Vkb), The user's location (Own), and A named location from a list, (Name).

- r. Select an option from the list. The option is displayed in the text box.
- s. Select the mandatory field Altitude from the Outline windowpane.
- t. Select the Altitude Virtual Keypad and enter the altitude or select the text box and enter the altitude with the keyboard.
- u. Select the Save As button. The Save As dialog box will be displayed.
- v. Select the Folder Text box and type the name for the folder that the message will be saved in and select the New Folder button. The new folder will be displayed in the Folders windowpane.

NOTE: If a folder was created earlier, then the user can highlight the folder and save the file to that folder.

- w. Select the File text box at the bottom of the Save As dialog box.
- x. Type the name for the message. The name will appear in the File text box.
- y. Select the OK button. The Save As Confirmation dialog box will be displayed.
- z. Select the OK button. The system will return to the Create: Airborne Artillery FCR Template.
- aa. Select the Send button. The Message Sent dialog box will appear stating "The message was sent " with an OK button.

NOTE: If the Default Message Addressing Settings or Message Addressing Settings within the message were not set prior to sending the message, the user would receive a Send Error dialog box saying No addressee selected with an OK button. The user would select the OK button and select the Message Addressing button and set the address.

- ab. Select the OK button. The system will return to the Create: Airborne Artillery FCR template.
- ac. Select the Close button. The system will return to the Ops screen.

2. Prepare and send a Personnel Status Report.
   - a. Select the Messages button on the function bar or the F4 button on the keyboard. The Messages dialog box will display.
   - b. Select the Create tab. The system will display the Create tab group.
   - c. Select the Reports radio button. The Report types will display in the Message type windowpane.
   - d. Select the Personnel Report. The option will be highlighted.
   - e. Select the Default Message Addressing button. The system will display the Message Addressing dialog box.
   - f. Select the Message Settings Tab.
   - g. Select the Precedence of the message. The proper radio button will highlight.
   - h. Select the desired acknowledge radio button. The proper radio button will highlight.
   - i. Select the Addresses Tab. The system will display the Addresses tab group.
   - j. Select the Select From: pull-down arrow. The system will display the three Address Books: Master, User, and Battalion.

STP 11-25C13-SM-TG

Performance Steps

    k. Select the Address Book that contains the Unit/Platform name that the message will be sent to. The Address Book will appear in the Select From: text box and all the units in that Address Book will appear in the Select From: window pane.

    l. Select the name of the Unit/Platform that the message needs to go to by selecting it from the Select From: windowpane on the left. The Unit/Platform name will be highlighted.

    m. Select the Add button. The Unit/Platform name will appear in the Address windowpane on the right side of the dialog box.

NOTE: If the user cannot find the Unit/Platform name, use the search button by selecting the search text box, typing the name, and selecting the Search button. If the name was spelled exactly the way it is in the UTO, then the name would highlight in the Select From: windowpane. The user can then select the Add button, which would place the Unit/Platform name in the Addresses windowpane. Do the same for all the unit/platform names that the message will go to.

    n. Select the OK button. The system will return to the Messages dialog box.

    o. Select the Execute button. The system will display the Personnel Status Report dialog box.

    p. Select the New button. The system will display the Add Personnel Record dialog box.

    q. Select the (Last Name): text box or the virtual keyboard and type the last name of the individual that the user is entering data for.

NOTE: Do the same for the First Name, Middle Name, Suffix, and SSN.

    r. Select the Nationality pull-down arrow. The system will display the available options.

    s. Select the Nationality of the individual. The system will display the Nationality in the text box.

NOTE: Do the same for Religion, Blood type, Unit Name, Role ID, Grade, MOS, and Status.

    t. Select the Male or Female radio button for the individual you are entering data for.

    u. Select the Apply button. The system will apply the data and present a new Add Personnel Record dialog box.

NOTE: Complete a record for each individual. After a record is done on each individual, select the Close button to get back to the Personnel Status Report dialog box.

    v. Select the SEARCH BY: pull down arrow at the bottom of the Personnel Status Report dialog box. The system will display the list of option that can be searched by.

    w. Select an option from the list. The option will display in the SEARCH BY text box.

NOTE: If the user selects Last Name, then the user will have to type the Last Name in the Search text box.

    x. Select the Search text box to the right of the SEARCH BY: pull down arrow. The cursor will blink inside the text box.

    y. Type the Last name of the individual that the user is looking for. The name will appear in the Search text box.

    z. Select the Search button. The system will highlight the record in the Personnel Status Report list at the top of the dialog box and activate the Modify button.

    aa. Select the Modify button. The system will display the Modify Personnel Record dialog box with the information on the individual being searched.

NOTE: The user can make changes to the record and apply them.

    ab. Select the Cancel button. The system will return to the Personnel Status Report dialog box.

    ac. Select the Close button. The system will return to the Ops screen.

3. Prepare and send a Freetext message.

    a. Select the Messages button on the function bar or the F4 button on the keyboard. The Messages dialog box will display.

    b. Select the Create tab. The system will display the Create tab group.

    c. Select the Reports radio button under the Msg Type field. The Reports radio button will highlight.

Performance Steps

    d. Scroll to the Freetext Message in the Msg Type windowpane. The message will highlight.
    e. Select the Execute button. The Create Freetext dialog box will display.
    f. Type a message in the text windowpane with the keyboard or the Virtual keyboard.

NOTE: The user can save the message before being sent or just send it without saving. The user can also change the address where the message will be sent by selecting the Message Addressing button.

    g. Select the Save As button. The system will display the Save As dialog box on top of the Create Freetext dialog box.
    h. Select a folder in the windowpane to save the message to by highlighting it or create a new folder by selecting the Folder text box and type a name for the new folder and select the Folder button. The folder will appear in the Folders windowpane.
    i. Highlight the new folder and select the File text box and type a name for the message that will be save to the folder.
    j. Select the OK button. The Save As Confirmation dialog box will appear stating "Message was saved" with an OK button.
    k. Select the OK button. The system will return to the Create Freetext dialog box.
    l. Select the Close button. The system will return to the Ops screen.

Performance Measures	**GO**	**NO GO**
1. Prepare and send an Airborne Artillery FCR Report.	——	——
2. Prepare and send a Personnel Status Report.	——	——
3. Prepare and send a Freetext message.	——	——

Evaluation Guidance: Score the Soldier GO if all steps are passed. Score the Soldier NO-GO if any step is failed. If the Soldier scores NO-GO, show him what was done wrong and how to do it correctly.

STP 11-25C13-SM-TG

PREPARE/SEND FIRE/ALERT MESSAGES USING FBCB2 VERSION 3.4
171-147-0009

Conditions: Given a vehicle with an operational Force XXI Battle Command Brigade and Below (FBCB2) system loaded with software version 3.4, you have received an operation order (OPORD) or fragmentary order (FRAGO) to conduct a tactical operation. Default Message Addressing settings have been set.

Standards: You must create and send a fire/alert message with significant and immediate information both horizontally and vertically on the battlefield. As a minimum, you must create and send: an Airborne Fire Mission, Subsequent Adjust, Check Fire, On Call Fire Cmd, and an End of Mission.

Performance Steps

NOTE: The Fire Support Coord Measures, MAYDAY Message, MOPP Alert, Observer Readiness Report, RECON Report, Threat Warning, and Strike Warning are done exactly the same way as the Airborne Fire Mission. The Call For Fire (CFF) is done exactly like task 171-147-0001 Step3 sub-step c. Subsequent Adjust, Check Fire, On Call Fire Cmd, and EOM messages are grayed out until a Fire mission is received from the Advanced Field Artillery Tactical Data System (AFATDS).

1. Prepare and send an Airborne Fire Mission.
    a. Select the Messages button on the Function Bar or the (F4) button on the Keyboard. The system will display the Message dialog box.
    b. Select the Create tab box.

NOTE: Default Transmission Settings should have been completed prior to PCC/PCIs. You cannot send or save a message until you have set your Default Transmission settings for each type of message or set the Transmission settings from within the message.

   c. Select the Fires/Alerts from the Msg Types field. The Fires/Alerts radio button will highlight.

NOTE: The Fires/Alerts function contains message templates for: Airborne Fire Mission, Call for Fire, Check Fire, EOM & Surveillance, Fire Support Coord Measures, Freetext Message, MAYDAY Message, MOPP Alert, Observer Readiness Report, On Call Fire Cmd, REDCON, and Strike Warning

   d. Select the Airborne Fire Mission. The System will highlight the Airborne Fire Mission.
   e. Select the Execute button. The system will display the Create: Airborne Fire Mission Template.
   f. Select the Outline Tab. The system will display the (M) Mandatory, (O) Optional, and (OG) Optional Group fields for this particular message.

NOTE: All of the fields with an M in front of them and are Red in color are mandatory fields and must be entered in order to be sent or saved. If you try to send or save the message before the mandatory fields are entered, you will get an error message saying, " The mandatory field Scan URN is not complete" with an OK button.

   g. Select the first mandatory field (Fire Mission Designator) by selecting the field name in the Outline windowpane. The option will be highlighted as well as the Fire Mission Designator in the right windowpane.
   h. Select the Fire Mission Designator pull-down arrow. The system will display a list of options.
   i. Select an option. The option will display in the text box.

NOTE: Once the mandatory fields are completed, you may send the message. You may also select optional fields such as Observer ID or Optional groups such as Target Location and enter the data. However, if you select an optional group, you will have to complete the mandatory fields from within that group.

   j. Select the Save As button. The system will display the Save As dialog box.
   k. Select the Folder text box. The cursor will blink in the text box.
   l. Type the Name you want to call the Folder. The name will be displayed in the text box
   m. Select the New Folder button. The new folder will be displayed in the Folder's windowpane.

STP 11-25C13-SM-TG

Performance Steps
  n. Select the Folder that you created by highlighting it.
  o. Select the File text box. The cursor will be blinking in the text box.
  p. Type the name that you want to call the message. It will be displayed in the text box
  q. Select the OK button. The Save As Confirmation dialog box will be displayed saying, "Message was saved."
  r. Select the OK button. The system will return to the Create: Airborne Fire Mission Template.
  s. Select the Send button. The system will display the Message Sent dialog box saying, "The message was sent."
  t. Select the OK button. The system will return to the Airborne Fire Mission Template.
  u. Select the Close button. The system will return to the Ops (MAP) screen.

NOTE: You must select the Messages button on the Function Bar or the F4 button on the keyboard and the Fires/Alerts radio button each time to get to the next Fires and alerts message.

  2. Perform a Subsequent Adjust.

NOTE: Subsequent Adjust Fire message is used to adjust fall of shot against an area target or for a registration fire mission. A Message to Observer (MTO) must be received by the users system from Advanced Field Artillery Tactical Data System (AFATDS) before the user can access the Subsequent Adjust, Check Fire, On Call Fire Cmd, and EOM tabs.

  a. Select the FIPR button. The FIPR dialog box will be displayed.
  b. Select the tab that has the MTO message.
  c. Highlight the MTO Message by selecting it from the list.
  d. Select the Display button at the bottom of the FIPR dialog box. The system will display the Combat Messages dialog box with Fire Mission Tab and Summary Sub-Tab selected.

NOTE: The Subsequent Adjust, Check Fire, On Call Fire Cmd, and EOM tabs are now active.

  e. Select the Subsequent Adjust Tab. The system will display the Subsequent Adjust Template.

NOTE: If the Subsequent Adjust template does not have all the desired options, the user can select the Long Form Message button at the bottom of the Combat Messages dialog box. This will bring up the Long Form template, which has more options. The Long Form is completed the same as all the Combat Messages.

  f. Select the Left or Right Radio button for the direction that the round needs to move. The radio button will highlight.
  g. Select the + button to increase or the - button to decrease the direction. The number will display in the text box.
  h. Select the Add or Drop Radio button for the distance the round needs to move. The Radio button will highlight.
  i. Select the + button to increase the distance or the - button to decrease the distance.
  j. Select the Up or Down radio button for the altitude. The Radio button will highlight.
  k. Select the + button to increase the height or the - button to decrease the height.
  l. Select the Method of Control pull-down arrow. The system will display the options.
  m. Select the desired option. The option will appear in the Text box.

NOTE: The Time On Target (Zulu) field will be Grayed out unless Timed Time on Target is selected in the Method of Control.

  n. Select the save button. The message stating "Message saved to file Subseq_DTG will be displayed in the Text window towards the bottom of the Combat messages dialog box.
  o. Select the send button. The message " Last Subsequent Adjust message sent at Date, Time Group (DTG)"
  p. Select the close button. The system will return to the FIPR button.
  q. Select the Close button. The system will return to the Ops screen.
  r. Select the FIPR button. The system will display the FIPR dialog box.

16 February 2005   3 - 139

STP 11-25C13-SM-TG

Performance Steps

    s. Select the Tab that has the MTO in it. The system will display the list of Messages under the Tab that you selected.
    t. Highlight the MTO message.
    u. Select the Display button. The system will display the Combat Messages dialog box with the Fire Mission Tab and the Summary Sub-Tab displayed.
    v. Select the Mission that you want to put the Check Fire on. The mission will be highlighted.
    w. Select the Check Fire Tab. The Check Fire Template will be displayed.
    x. Select the proper Check Fire Radio button for the action that you want to do.

NOTE: Check Fire Order is for Checking a certain mission, Check Fire All is for checking all missions, and Cancel Check Fire is for canceling the Check Fire command.

    y. Select the send button. The system will display "Last Check Fire Message sent at DTG" in the Message window.
    z. Select the Close button. The system will return to the FIPR dialog box.
    aa. Select the Close button. The system will return to the Ops screen.

3. Perform the Check Fire Function.

NOTE: Remember that an MTO must be received by the users system from AFATDS to get access to this tab.

    a. Select the FIPR button. The FIPR dialog box will be displayed.
    b. Select the tab that has the MTO message.
    c. Highlight the MTO Message by selecting it from the list.
    d. Select the Display button at the bottom of the FIPR dialog box. The system will display the Combat Messages dialog box with Fire Mission Tab and Summary Sub-Tab selected.
    e. Select the Check Fire tab. The system will display the Check Fire tab group.
    f. Select the radio button for the desired option under the Check Fire/Cancel Check Fire Cmd field. The radio button will highlight.
    g. Select the Send button. The system will display "Message sent at DTG" in the message pane.
    h. Select the Close button. The system will return to the FIPR Queue.

4. Perform the On Call Fire Cmd function.

NOTE: The On Call Fire Command " message is used by friendly units to send Fire commands to the Fire support unit for the Fire mission selected from the Summary tag group. An MTO must be received before the "On Call Fire Cmd" tab can be activated.

    a. Select the FIPR button. The system will display the FIPR dialog box.
    b. Select the Tab that has the mission that you want to call the indirect fires on.
    c. Select the Mission by highlighting the Originator and DTG of the mission.
    d. Select the display button. The system will display the Combat Messages dialog box with the Fire Mission Tab and the Summary Sub-tab selected.
    e. Select the On call Fire Cmd Tab. The On Call Fire Cmd Template will be displayed with a message saying, " An "FO Command" to fire the mission will be transmitted by sending this message!"
    f. Select the Send button if the user wants the mission to fire. The message " Last On Call Fire Cmd message sent at DTG" will be displayed in the message window pane Or select the Cancel button to the right of the message window.
    g. Select the Close button. The system will return to the FIPR dialog box.
    h. Select the Close button. The system will return to the Ops screen.

5. Perform the End of Mission (EOM) function.

NOTE: The End of Mission function is used to direct the end of mission processing of a fire mission, selected from the Summary tab group and to provide target surveillance and provide Fire mission refinement data.

    a. Select the FIPR button. The system will display the FIPR dialog box.

Performance Steps
    b. Select the Tab that has the mission that the user wants to associate the EOM to.
    c. Select the mission from the list by highlighting it.
    d. Select the Display button. The system will display the Combat messages dialog box with the Fire Mission Tab and the Summary Sub-tab.
    e. Select the EOM Tab. The system will display the EOM Template.
    f. Select the EOM Type pull-down arrow. The system will display End of Mission and End of Mission & Surveillance.
    g. Select the desired option from the list. The option will be displayed in the text box.

NOTE: The Surveillance Info field will be grayed out unless the user selects the End of Mission & Surveillance option from the EOM Type field.

    h. Select the Effect Achieved: pull-down arrow. The system will display the options list.
    i. Select the desired option from the list. The option will be displayed in the text box.
    j. Select the + button to increase or the - button to decrease the number of Enemy Casualties.
    k. Select the send button. The message "Last EOM message sent at DTG"
    l. Select the Close button. The system will return to the FIPR dialog box.
    m. Select the Close button. The system will return to the Ops screen.

Performance Measures	GO	NO GO
1. Prepare and send an Airborne Fire Mission.	——	——
2. Perform a Subsequent Adjust.	——	——
3. Perform the Check Fire Function.	——	——
4. Perform the On Call Fire Cmd function.	——	——
5. Perform the End of Mission (EOM) function.	——	——

Evaluation Guidance: Score the Soldier GO if all steps are passed. Score the Soldier NO-GO if any step is failed. If the Soldier scores NO-GO, show him what was done wrong and how to do it correctly.

STP 11-25C13-SM-TG

PREPARE/SEND ORDER/REQUEST MESSAGES USING FBCB2 VERSION 3.4
171-147-0010

Conditions: Given a vehicle with an operational Force XXI Battle Command Brigade and Below (FBCB2) system loaded with software version 3.4, you have received an operation order (OPORD) to conduct a tactical operation. Default Message Addressing settings have been set.

Standards: You must create and send an Orders/Requests message with significant and immediate information both horizontally and vertically on the battlefield. As a minimum, you must create and send: a Fragmentary Order, Freetext Message, MEDEVAC Request, CTIL Action, and a LOG Call for Support.

Performance Steps

NOTE: The LOG Call for Support, Operation Order, Operations Plan, Unit Ref Query, and the Warning Order are completed the same way as the Fragmentary Order. For each message type that the user wants to complete, the user will need to select the Messages button, Create tab, Orders/Requests radio button, and then the Message type for each message.

1. Prepare and send a Fragmentary Order.

NOTE: Fragmentary Order is an abbreviated version of the Operations Order and is used by the Commander / Staff to issue plans/orders to subordinate units to effect the coordinated execution of an operation.

    a. Select the Messages button on the Function Bar or the (F4) button on the Keyboard. The system displays the Messages dialog box.

NOTE: The Messages function displays six tabs: Send, Create, Edit, Print, Manage, and Sent Queue.

    b. Select the Create tab. The system will display the Create tab group.

NOTE: All the Tabs in the Messages dialog box have three common push buttons at the bottom of the Tab which consists of: Execute, which activates the selection, Close which closes the dialog box, and Help which gives help on that tab.

    c. Select the Orders/Requests radio button from the Msg Type Field. The radio button will highlight.
    d. Select the Fragmentary Order from the Message windowpane. The Fragmentary Order will be highlighted.

NOTE: The Default Transmission Settings should be completed prior to the PCC/PCIs. You cannot send or save a message until you have set the Default Transmission Settings for each type of message or the Transmission Settings from within the Message.

    e. Select the Execute button. The system will display the Create: FRAGO Template.
    f. Select the Outline tab. The system will display the (M) Mandatory, (O) Optional, (OG) Optional Groups fields.

NOTE: All the Fields with the M in front of them and are red in color must be completed in order for the message to be sent or saved. If you try to send or save the file without entering data into the mandatory fields, an error message saying " The Mandatory field Scan URN is not complete" with an OK button. If this happens, select the OK button and go back and enter the Mandatory fields that were missed.

    g. Select the mandatory field (OPORD Name) from the Outline windowpane. The option will be highlighted.
    h. Select the OPORD Name text box from the window on the right of the Create: FRAGO Template. The OPORD Name text box will activate and the cursor will blink.
    i. Type the Name that you want to call the OPORD. The Name will be displayed in the text box.

Performance Steps

NOTE: Once the mandatory fields are completed, you may send the Order or Request. You may also select optional fields like FRAGO ID Number and optional groups like Annex Data (0/20) and enter them. However, if you select an optional group, you will need to enter the mandatory fields from within that group.

   j. Select the Save As button. The system will display the Save As dialog box.
   k. Select the Folder text box and type the Folder Name and select the New Folder button or select a folder that was already created by highlighting the folder name in the folders window.
   l. Select the File text box and type the name that you want to call the Order or Request.
   m. Select the Ok button. The system will display the Save As Confirmation dialog box saying, "Message was saved" with the OK button.
   n. Select the OK button. The system will return to the Create: FRAGO Template.
   o. Select the Send button. The message "The Message was sent" with an OK button will be displayed.

NOTE: If the Default Transmission Settings or Transmission Settings were not set prior to sending this Order or request, you would receive an Error message saying, "No addressee selected" with an OK button. You would have to select the OK button and select the Transmission Settings button and set the address that you want the message to go to.

   p. Select the Close button. The system will return to the Ops Screen.

2. Prepare and send a Freetext Message.

NOTE: A Freetext Message is used to send information that is not covered in the other message types in FBCB2. It also is used to send a plain text message.

   a. Select the Create tab. The Create tab fields will display.
   b. Select the Orders/Requests radio button. The radio button will highlight.
   c. Select the Freetext Message from the Message type windowpane. The message will be highlighted.
   d. Select the Set Default Transmission Settings button. The Transmission Settings dialog box will be displayed.
   e. Select the Settings Tab.
   f. Enter the Precedence, Acknowledgement, Retries, and Perishability DTG.
   g. Select the Add Addresses tab.
   h. Select the Address that you want to send the message to and add it to the right windowpane as an action or info address.
   i. Select the OK button. The Apply dialog box will display.
   j. Select the OK button. The system will return back to the Message dialog box with the Freetext Message still highlighted.
   k. Select the Execute button. The system will display the Create: Freetext dialog box.
   l. Type the message that you want to send in the text windowpane.
   m. Select the Save As button. The Save As dialog box will be displayed.
   n. Select the Folder from the Folders window that you want to save the message to or Create a new folder by selecting the Folder text box and typing the name that you want to call it and selecting the New Folder button. This will create a new folder in the Folders windowpane.
   o. Highlight the new folder in the Folders windowpane.
   p. Select the File text box and type the name that you want to call the message.
   q. Select the OK button. The Save As dialog box will be displayed saying, "Message was saved." with an OK button.
   r. Select the OK button. The system will return to the Create: Freetext dialog box with the new name at the top.
   s. Select the Send button. The system will display the Message Sent dialog box.
   t. Select the OK button. The system will return to the Create: Freetext dialog box.
   u. Select the Close button. The system will return to the Ops screen.

STP 11-25C13-SM-TG

Performance Steps

3. Prepare and send a MEDEVAC Request.

NOTE: Use the MEDEVAC request message to request ground or aircraft support to evacuate friendly and/or enemy casualties. The MEDAVAC tab group is accessed from Combat Msgs button on the Function Bar or the F3 button on the keyboard, Or from the Messages button under Orders/Request.

NOTE: The MEDEVAC Request is also addressed as part of the task 171-146-0001 "Prepare/Send Combat Messages Using FBCB2 Version 3.3."

   a. Select the Messages button on the Function Bar or the F4 button on the keyboard.
   b. Select the MEDAVAC Request. The system will highlight the MEDAVAC Request.
   c. INSERT DEFAULT TRANS
   d. Select the Execute button. The system will display the Combat Messages dialog box.
   e. Select the Fill Loc button. The Combat Messages dialog box will disappear and the mouse cursor will be replaced with a cross hair.
   f. Select a location on the map where the pick up site will be. The system displays the location in the Location text box.
   g. Or Select the Fill LRF button. The system will input the location provided by the laser range finder.

NOTE: This function works if your vehicle is equipped with the Laser Range finder.

   h. Select the Amb Patients + or - button to increase or decrease the number of Ambulatory patients.
   i. Select the Ltr Patients + or - button to increase or decrease the number of Litter Patients.
   j. Select the Marking pull-down arrow. The system will display the Marking options list. Select an option from the list. The system displays the selection in the Marking text box.
   k. Select the Color pull-down arrow. The system will display the Color options list. Select an option from the list. The system displays the selection in the Color text box.
   l. Select the Pickup Zone Hot check box. The system will display a check mark in the Pickup Zone Hot text box.

NOTE: If the area is not Hot, then do not check this box. Hot meaning that the area has hostiles.

   m. Select the MEDAVAC Priority pull-down arrow. The system will display the MEDAVAC Priority options list. Select and option from the list. The system will display the selection in the MEDEVAC Priority text box.
   n. Select the NBC Contamination pull-down arrow. The system will display the NBC Contamination option list. Make a selection

4. Prepare and send a CTIL Action.

NOTE: The CTIL Action message allows the authorized user (Access Level 4) to modify the Commander's Tracked Items List (CTIL) to meet the Unit's specific requirement's. Upon receipt and acceptance by the receiver, this message overwrites the previously received CTIL. FBCB2 user should validate the message sender prior to accepting the new CTIL message. If an errant CTIL message is loaded, the only way to replace it is to call for another CTIL message from the proper logistic source.

   a. Select the Messages button on the Function Bar, or the F4 button on the keyboard. The Messages dialog box will display.
   b. Select the Create tab. The system will display the Create tab group.
   c. Select the Orders/Request radio button from the Msg Type field. The button will highlight.
   d. Select the CTIL Action message by highlighting it. The message will highlight.
   e. Select the Execute button. The system will display the Using Default CTILS dialog box.
   f. Place a checkmark in the box to the left of each item that is specific to the unit. A checkmark will display in each box that was selected.

STP 11-25C13-SM-TG

Performance Steps

 g. Select the Message Addressing button. The system will display the Message Addressing dialog box.

NOTE: If the Default Message Addressing settings for the CTIL Action message were completed prior to preparing this message, then the Message addressing can be skipped.

 h. Select the Message Settings tab. The Message Settings tab group will display.
 i. Select the desired Precedence. The selected radio button will highlight.
 j. Select the desired Acknowledge option. A checkmark will be displayed in the selected box.

NOTE: Some Acknowledge options will be grayed out depending on the selected message.

 k. Select the proper Security Level.

NOTE: The Security level can only be changed by authorized personnel only. The Security Level may be grayed out.

 l. Select the DTG button under the Perishability DTG: field. The DTG Keypad dialog box will display.
 m. Enter the proper DTG by selecting the + or - buttons to increase or decrease the number. The numbers will display in the text boxes.
 n. Select the OK button. The system will return to the Message Addressing dialog box.
 o. Select the Addresses tab. The system will display the Addresses tab group.
 p. Select the Selected From: pulldown arrow. The system will display the three types of users groups.
 q. Select the desired user's group. The selected user group will display in the Selected From: text box with a list of the units in the user's group.
 r. Select the unit by either highlighting it in the Selected From: windowpane, or selecting the Search text box and typing the name of the unit and selecting the search button. The unit will highlight in the Selected From: windowpane.
 s. Select the Add button. The unit will display in the right window pane under the Addresses tab.

NOTE: Do the same for all the Unit/Platforms that the user wants the message to go to.

 t. Select the OK button. The system will return to the Using Default CTILS dialog box.
 u. Select the Send CTILs button. The system will display a CTILs Sent dialog box.
 v. Select the OK button. The system will return to the Using Default CTILS dialog box.
 w. Select the Close button. The system will return to the Ops screen.

5. Create and send a LOG Call for Support.

NOTE: The LOG Call for Support is threaded messages that the operator can send to request logistic support. Typically a call for support is sent through the chain of command to the supporting unit using the LOG Call for Support message "Orders/Requests."

 a. Select the Messages button on the Function Bar, or the F4 button on the keyboard.
 b. Select the Create tab. The system will display the Create tab group.
 c. Select the Orders/Requests radio button. The radio button will highlight.
 d. Select the LOG Call for Support message. The message will highlight.
 e. Select the Execute button. The Create: LOG Call for Support dialog box will display.

NOTE: The fields with red fonts with an "M" in front of them under the Outline tab are the mandatory fields that must be entered in order to send the message. The message will not send unless these fields are completed. Unit SOPs may require more fields to be completed. There is a short cut the user can use to quickly get to the field to enter data. By selecting the field in the Outline tab, it will highlight the text box for the field that was selected. The user can now enter the data for that field.

 f. Select the "Request Type pull down arrow. The system will display the available options.
 g. Select the desired option. The option will display in the text box.
 h. Select the Action pull down arrow. The system will display the available options.

STP 11-25C13-SM-TG

Performance Steps
   i. Select the desired option. The option will be displayed in the text box.
   j. Select the Supporting Unit pull down arrow. The system displays the Unit Name dialog box.
   k. Select the Name of the Unit that needs the support. The Unit Name will display in the text box.
   l. Select the Point of Contact pull down arrow. The Unit Name dialog box will display.
   m. Select the Name of the Unit that was selected in the Supporting Unit text box. The Unit name will display in the text box.
   n. Select the Mission Location pull down arrow. The available options will display.
   o. Select the desired option. The option will display in the text box.

NOTE: If available options are: Map, which is selecting a location on the map; (LRF), which is using a Laser Range Finder; (Kbd), which is using the Keyboard to enter the location; (Vkb), which is using the Virtual Keyboard; Own, which is selecting your own location; and Name, which is selecting from a list of names.

   p. Scroll down to the Comments field and select the Edit/View button. The system will display the Edit/View Text dialog box.
   q. Type the comments that will transmit with the message. The comment will display in the text box.
   r. Select the OK button. The system will return to the Create: LOG Call for Support dialog box.
   s. Select the Save As button. The system will display the Save As dialog box.
   t. Select the Folder that the message will be saved to, or create a new folder and then highlight the folder. The folder will be highlighted.
   u. Select the File text box and type the name that the user wants to call the message.
   v. Select the OK button. The system will display the Save AS Confirmation dialog box with an OK button.
   w. Select the OK button. The system will return to the Create: LOG Call for Support dialog box.

NOTE: The user can send the message if the Message Addressing settings were completed earlier or the user can select the Message Addressing button and enter the address the same way it was done earlier in the task.

   x. Select the Send button. The system will display the Message Sent dialog box with an OK button.
   y. Select the OK button. The system will return to the Create: LOG Call for Support dialog box.
   z. Select the Close button. The system will return to the Ops screen.

Performance Measures	**GO**	**NO GO**
1. Prepare and send a Fragmentary Order.	___	___
2. Perform a Freetext Message.	___	___
3. Prepare and send a MEDEVAC Request.	___	___
4. Prepare and send a CTIL Action.	___	___
5. Create and send a LOG Call for Support.	___	___

Evaluation Guidance: Score the Soldier GO if all steps are passed. Score the Soldier NO-GO if any step is failed. If the Soldier scores NO-GO, show him what was done wrong and how to do it correctly.

STP 11-25C13-SM-TG

PERFORM BEFORE-OPERATIONS PREVENTIVE MAINTENANCE CHECKS AND SERVICES ON FBCB2 VERSION 3.4
171-147-0011

Conditions: Given a vehicle with an operational Force XXI Battle Command Brigade and Below (FBCB2) system with software version 3.4 loaded and TM 11-7010-326-10.

Standards: As a minimum you must: Make sure that the vehicle's MASTER POWER has been turned off. Check the following components for damage: AN/UYK-128 (V) Computer, Display Unit, Processor unit, Keyboard unit, and Cables and Connectors. Notify unit maintenance of any damaged or missing parts. Follow all warnings and cautions in the TM to prevent injury to personnel and or damage to equipment.

Performance Steps

NOTE: The Appliqué+ V4 computer requires daily and weekly checks. Daily checks are to be performed before operation, during operation, and after operation as indicated in the "Service Interval" column.

1. Check the AN/UYK-128(V) Computer.
   a. Ensure all computer components, accessories, and cables are present, secured and properly stowed to avoid damage.
   b. Check mounting bolts before extensive equipment use.
   c. Check the ram ball mount assembly for tightness, if equipped.

NOTE: If the processor unit (PU), display unit (DU), keyboard unit (KU), or connecting cables are missing/damaged, they are not fully mission capable. The system is degraded if: PLGR, SINCGARS, or EPLRS (if so equipped) are missing/damaged.

2. Check the Touchscreen on display unit (DU).
   a. Check for cracks, severe scratches, or other damage to the screen.

NOTE: The touchscreen is degraded if: Touchscreen is cracked or has severe scratches that would prevent its proper operation.

3. Check the Grounding strap on display unit (DU) chassis.
   a. Conduct visual check to see if the grounding strap is present and tight on the bottom of the DU.

NOTE: The equipment is not fully mission capable if the grounding strap for the DU chassis is missing, frayed, or disconnected.

4. Check the processor unit (PU).
   a. Check the battery tray/battery box inside the PU for charge.

NOTE: The Processor Unit is degraded if: Battery charge indicators show that battery(s) have a low charge/ no charge. For PU with NSN 7025-01-474-3793, press button twice on front of battery box to display diagnostic codes. Ensure that code "05" (i.e., Back-up Battery Low) is not displayed. For PU with NSN 7025-01-475-0217, check front of battery tray to ensure that the charge indicator displays at least three (3) LCD bars.

5. Check the Grounding strap on Processor Unit (PU) chassis.
   a. Check to make sure that the Grounding strap is present, not frayed or disconnected.

NOTE: The Processor Unit is not mission capable if: the Grounding strap is missing, frayed or disconnected or the Grounding strap cannot be fastened securely

6. Check the keyboard unit (KU).
   a. Check KU for non-functioning/ missing alphanumeric or Enter keys.

Performance Steps

NOTE: The keyboard unit is not mission capable if it has missing keys or alphanumeric, or Enter keys that do not function.

7. Check the cables and connectors.
    a. Check for cables with frayed, broken, or bare wires.
    b. Check to see that all cable connectors are properly mated (i.e., only the blue band should be visible).

NOTE: Cables and Connectors are not mission capable if: Wires or cables are damaged (i.e., frayed, broken, or bare). Connector(s) are not properly mated and any red band is visible on connector(s).

Performance Measures	GO	NO GO
1. Check the AN/UYK-128(V) computer.	——	——
2. Check the touchscreen on display unit (DU).	——	——
3. Check the grounding strap on display unit (DU) chassis.	——	——
4. Check the processor unit (PU).	——	——
5. Check the grounding strap on processor unit (PU) chassis.	——	——
6. Check the keyboard unit (KU).	——	——
7. Check the cables and connectors.	——	——

Evaluation Guidance: Score the Soldier GO if all steps are passed. Score the Soldier NO-GO if any step is failed. If the Soldier scores NO-GO, show him what was done wrong and how to do it correctly.

STP 11-25C13-SM-TG

PERFORM SHUTDOWN PROCEDURES FOR FBCB2 VERSION 3.4
171-147-0012

Conditions: Given a vehicle with an operational Force XXI Battle Command Brigade and Below (FBCB2) system loaded with software version 3.4, TM 11-7010-326-10, Precision Lightweight Global Positioning System Receiver (PLGR), enhanced position location reporting system (EPLRS) if equipped, and Single-channel Ground and Airborne Radio System (SINCGARS) with Internet controller (INC).

Standards: As a minimum you must: Shut down the FBCB2 computer, PLGR, SINCGARS, and EPLRS. Perform all steps in sequence. Follow the warnings and cautions in the TM to prevent injury to personnel and or damage to equipment.

Performance Steps

NOTE: This task describes shutdown procedures for the AN/UYK-128(V) computer. The AN/UYK-128(V) computer must be shutdown first before the PLGR, EPLRS (if equipped), and SINCGARS ASIP (if equipped).

CAUTION: Do not shutdown power to the computer without first following software shutdown procedures. Failure to comply may cause the loss of program data.

CAUTION: Leaving the Processor Unit (PU) circuit breaker/switch set to "ON" will enable the battery pack to continuously charge as long as there is 18-33 Volts Direct Current (VDC) power available. This could possibly result in a dead vehicle battery if left in this condition over an extended period.

CAUTION: The keyboard should be disconnected and properly stowed when not in use to prevent it from causing equipment damage.

1. Perform AN/UYK-128(V) computer shutdown.
    a. Select the F6 Admin button. The Admin dialog box is displayed.
    b. Select the Exit Ops button. The Exit Ops confirmation dialog box is displayed.
    c. Select the Yes button. Exit Ops confirmation dialog box closes. The Ops Auto-Log in dialog box opens with countdown timer started.
    d. Select Cancel Time out button. The Ops Auto-Log in dialog box closes.
    e. Select the Start button. The Start button option menu is displayed.
    f. Select the Shut Down option. The Shut down option menu is displayed.
    g. Select the Shutdown option. Shut down confirmation dialog box is displayed
    h. Select the Yes button. Screen displays: Shutting Down the System. Safe to power off when the screen message says, Type any key to continue or approximately 10 seconds after syncing file systems...done is displayed.
    i. Press DU PWR button for up to 4 seconds and release after DU PWR LED goes dark.
    j. Set the circuit breaker/switch on the PU to the "OFF" position. The circuit breaker/switch is pointed toward the outside edge of the PU.
    k. Ensure system is properly secured (e.g., PU, DU and KU locked and secured).

2. Perform PLGR shutdown.
    a. Press the OFF (0) button for two seconds. PLGR displays: Unit Turning OFF in __ seconds ON: to cancel OFF: quick off.

3. Perform SINCGARS ASIP shutdown.
    a. Place the function switch to STBY or OFF per SOP. The SINCGARS display goes blank (i.e., dark).

4. Perform INC shutdown.
    a. At the VAA set the CB1 POWER toggle switch to OFF (i.e., down position). The DS1 green POWER light goes off.

16 February 2005

STP 11-25C13-SM-TG

Performance Steps

    5. Perform EPLRS shutdown.
        a. Turn POWER switch to the OFF position. Green POWER indicator LED goes off.

Performance Measures	**GO**	**NO GO**
1. Perform AN/UYK-128(V) computer shutdown.	____	____
2. Perform PLGR shutdown.	____	____
3. Perform SINCGARS ASIP shutdown.	____	____
4. Perform INC shutdown.	____	____
5. Perform EPLRS shutdown.	____	____

Evaluation Guidance: Score the Soldier GO if all steps are passed. Score the Soldier NO-GO if any step is failed. If the Soldier scores NO-GO, show him what was done wrong and how to do it correctly.

STP 11-25C13-SM-TG

PERFORM DURING-OPERATIONS PREVENTIVE MAINTENANCE CHECKS AND SERVICES ON FBCB2 VERSION 3.4
171-147-0013

Conditions: Given a vehicle with an operational Force XXI Battle Command Brigade and Below (FBCB2) system with software version 3.4 loaded and TM 11-7010-326-10. The Before checks have been completed.

Standards: As a minimum you must: Perform the During Preventive Maintenance Checks and Services on the FBCB2 system in sequence with the TM. The Before checks have been completed.

Performance Steps

1. Verify that the green LEDs are illuminated for PWR (Power), DISP (Display), and CPU (Processor Unit) on the DU Controls and Indicators Panel.

NOTE: Equipment is not fully mission capable if: Any red LED remains continuously lit when operating. Shutdown the AN/UYK-128(V) immediately. Perform Troubleshooting procedures. The red and/or amber LEDs may Illuminate briefly during initial power application.

2. Verify that the Local Comm status is "G" (green) which is the first gumball from the left on the Classification/Status Bar on the FBCB2 Ops Main screen.

NOTE: The equipment is degraded if: Local Comm status gumball is "A" (Amber). Status is Unknown if gumball is "W" (White).

3. Verify that the Global Positioning System (GPS) status is "G" (green) on the FBCB2 Ops Main screen, which is the second gumball from the left on the Classification/Status Bar.

NOTE: The equipment is degraded if: GPS status gumball is "R" (red) or "A" (amber). Status is Unknown if gumball is "W" (white).

4. If so equipped, verify that the Battlefield Combat Identification System (BCIS) status is "G" (green) on the FBCB2 Ops Main screen.

NOTE: The Equipment is degraded if: BCIS status gumball is "R" (red) or "A" (amber), which is the third gumball from the left on the Classification/Status Bar. Status is Unknown if gumball is "W" (white).

Performance Measures	GO	NO GO
1. Verify that the green LEDs are illuminated for PWR (Power), DISP (Display), and CPU (Processor Unit) on the DU Controls and Indicators Panel.	——	——
2. Verify that the Local Comm status is "G" (green) which is the first gumball from the left on the Classification/Status Bar on the FBCB2 Ops Main screen.	——	——
3. Verify that the Global Positioning System (GPS) status is "G" (green) on the FBCB2 Ops Main screen, which is the second gumball from the left on the Classification/Status Bar.	——	——
4. If so equipped, verify that the Battlefield Combat Identification System (BCIS) status is "G" (green) on the FBCB2 Ops Main screen.	——	——

Evaluation Guidance: Score the Soldier GO if all steps are passed. Score the Soldier NO-GO if any step is failed. If the Soldier scores NO-GO, show him what was done wrong and how to do it correctly.

STP 11-25C13-SM-TG

PERFORM AFTER-OPERATIONS PREVENTIVE MAINTENANCE CHECKS AND SERVICES ON FBCB2 VERSION 3.4
171-147-0014

Conditions: Given a vehicle with an operational Force XXI Battle Command Brigade and Below (FBCB2) system with software version 3.4 loaded and TM 11-7010-326-10. The Before and During checks have been completed.

Standards: As a minimum you must: Perform the After Preventive Maintenance Checks and Services on the FBCB2 system in sequence with the TM. The Before and During checks have been completed.

Performance Steps

1. Check the AN/UYK-128 (V) Computer.
   a. Ensure all computer components, accessories, and cables are present, secured and properly stowed to avoid damage.
   b. Check mounting bolts before extensive equipment use.
   c. Check Ram Ball assembly for tightness, if equipped.

NOTE: Equipment is degraded if: PLGR, SINCGARS, or EPLRS (if so equipped) are missing/damaged and if Processor Unit (PU), Display Unit (DU), Keyboard Unit (KU), or connecting cables are missing/damaged, and thereby prevent proper equipment operation.

2. Check the Processor Unit (PU).
   a. Securely fasten the Removable Hard Disk Cartridge (RHDDC) access door. All four (or all six) captive fasteners must be evenly and securely tightened.

NOTE: Equipment is not fully mission capable if: RHDDC access door is not properly closed due to obstructions, bad seal, and broken/missing/loose captive fastener(s).

3. Check the Processor Unit (PU), Display Unit (DU), and Keyboard Unit (KU).
   a. Safeguard with 5200 series lock(s)/cables to secure PU, DU, and KU as applicable to your platform.

NOTE: Equipment is not fully mission capable if: PU, Hard Disk Drive Door, DU, and/or KU are not secured and cables and/or locks are missing/damaged.

4. Set Circuit Breaker (CB) toggle switch to OFF position.

NOTE: Equipment is degraded if: CB toggle switch is not set to OFF position or CB switch is broken/damaged.

Performance Measures	GO	NO GO
1. Check the AN/UYK-128 (V) Computer.	——	——
2. Check the Processor Unit (PU).	——	——
3. Check the Processor Unit (PU), Display Unit (DU), and Keyboard Unit (KU).	——	——
4. Set Circuit Breaker (CB) toggle switch to "OFF" position.	——	——

Evaluation Guidance: Score the Soldier GO if all steps are passed. Score the Soldier NO-GO if any step is failed. If the Soldier scores NO-GO, show him what was done wrong and how to do it correctly.

PREPARE/SEND A LOGISTICAL STATUS REPORT USING FBCB2 VERSION 3.4
171-147-0015

Conditions: Given a vehicle with an operational Force XXI Battle Command Brigade and Below (FBCB2) system with software version 3.4 loaded. You are in an assembly area or battle position and you have received a platoon operation order (OPORD) or fragmentary order (FRAGO) for a resupply operation. The default Message Addressing settings for this report have been set.

Standards: As a minimum you must: Select the LOG Report on FBCB2 and enter your logistical data in the on-hand column. Save the updated information and send the report to higher.

Performance Steps

1. Select the Apps button from the Function bar or select the F7 button from the keyboard. The system will display the Apps dialog box.

2. Select the FBCB2 tab which should be the Defaulted Tab. The system will display the FBCB2 tab group.

3. Select LOG Report under the FBCB2 Tab. The system will highlight the LOG Report.

NOTE: The LOG Report can also be accessed through the Messages button, Create tab, and Reports radio button. The Logistics Status Report message is used to report an individual unit's, multiple units', or a compilation of the units' commander's tracked items list (CTIL) and basic required items list (BRIL) on-hand quantities. FBCB2 receives LOGSITREP data from other single FBCB2 platforms. FBCB2 will not automatically save multiple LOGSITREPs associated with the same unit or operational facility (OPFAC); new reports will overwrite old ones. If the logistics application is active when a "Logistics" Report message arrives, a flag icon appears on the [Redisplay] button highlighted in yellow.

4. Select the Execute button. The system will display the LOG Report dialog box with a Roll up tab and a Single tab.

NOTE: The Roll up tab is divided into three areas: (1) The operational Information, which includes the Reporting DTG, Roll up Unit, Quantity Displayed, Roll up Check box, Roll up Comments button, and the Subordinate Unit Information. (2) The CTIL Data display, which includes the Individual items of the CTIL and the Class of supply on the left side. (3) The push buttons, which include the Roll up, Message Addressing, Tailor CTILs, Select All, Send, Redisplay, Deselect All, Save, Close, Delete, Print, and Help. There will be from two columns to several columns on the Roll up tab depending on the user's duty position. The first column is the Roll up column, which is the column below the Roll up Comments button. The user's column is the column just to the right of the Roll up column. If subordinate platforms will be sending their LOG Report to this platform, then there will be several more columns to the right of the user's column.

5. Select the Single tab. The system displays the Single tab group.

NOTE: The Single Tab is similar to the Roll up tab. The Single tab is the tab that the quantities will be entered into.

6. Select the text box under the On-Hand/Operational Item Count column for the first item. The system will display NS highlighted in black.

7. Enter the amount that the user has on-hand in the text box. The system will display the amount in the text box.

NOTE: If the number will not enter in the text box, then go back to he Roll up Tab and uncheck the User's column, then go back to the Single tab and enter the number.

8. Select the text box under the Authorized Item Count column. The system will display the NS highlighted in black.

STP 11-25C13-SM-TG

Performance Steps

9. Enter the amount that the user is authorized. The system will display the amount in the text box.

10. Place the cursor in the text box under the Required Item Count Column for the first item and click the left mouse button. The system will automatically calculate the difference.

11. Select the Save button. The system will display a Successful Save dialog box stating "Saved the logistics record for (the user's platform name)" with an OK button.

12. Select the OK button. The system will return to the Single tab.

NOTE: Complete steps 6 through 12 for each item in the CTIL.

13. Select the Roll up Tab. The system will display the Roll up tab group with the quantities entered in the Single tab.

14. Place a checkmark in the checkbox above the user's column next to the word Selected and any column to the right of the user's column.

NOTE: Before selecting the Roll up button, ensure that the Selected Checkbox above the Roll up column is check marked and all the columns to the right of the user's column to include the user's column are check marked.

15. Select the Roll up button. The system will roll up all the quantities for all the columns into the Roll up column.

16. Select the Roll up Comments button. The system will display the Comments: dialog box.

17. Type a comment that the user wants to send with the keyboard or the virtual keyboard.

18. Select the OK button. The system will return to the Roll up tab with a checkmark on the Roll up Comments button.

NOTE: If the LOG Report needs to be sent higher, then the user would check the Redisplay button if there was a Black flag highlighted in yellow to update the Log Report, Roll up the new data and send it to higher.

19. Select the Close button. The system will return to the Ops screen.

Performance Measures	**GO**	**NO GO**
1. Select the Apps button from the Function bar or select the F7 button from the keyboard.	——	——
2. Select the FBCB2 tab which should be the Defaulted Tab.	——	——
3. Select LOG Report under the FBCB2 Tab.	——	——
4. Select the Execute button.	——	——
5. Select the Single tab.	——	——
6. Select the text box under the On-Hand/Operational Item Count column for the first item.	——	——
7. Enter the amount that the user has on-hand in the text box.	——	——
8. Select the text box under the Authorized Item Count column.	——	——
9. Enter the amount that the user is authorized.	——	——
10. Place the cursor in the text box under the Required Item Count Column for the first item and click the left mouse button.	——	——

Performance Measures	**GO**	**NO GO**
11. Select the Save button.	——	——
12. Select the OK button.	——	——
13. Select the Rollup Tab.	——	——
14. Place a checkmark in the checkbox above the user's column next to the word Selected and any column to the right of the user's column.	——	——
15. Select the Rollup button.	——	——
16. Select the Rollup Comments button.	——	——
17. Type a comment that the user wants to send with the keyboard or the virtual keyboard.	——	——
18. Select the OK button.	——	——
19. Select the Close button.	——	——

Evaluation Guidance: Score the Soldier GO if all steps are passed. Score the Soldier NO-GO if any step is failed. If the Soldier scores NO-GO, show him what was done wrong and how to do it correctly.

STP 11-25C13-SM-TG

EMPLOY MAP FUNCTIONS USING FBCB2 VERSION 3.4
171-147-0017

Conditions: Given a vehicle with an operational Force XXI Battle Command Brigade and Below (FBCB2) system with software version 3.4 loaded, precision lightweight global positioning system receiver (PLGR), enhanced position location radio system (EPLRS), and single-channel ground air radio system (SINCGARS) with Internet controller (INC).

Standards: As a minimum you must: Employ Map functions to manage and exchange information, to include: Background type, Map scale, Zoom, Brightness, Contrast, Grid Coordinate type, MGRS accuracy, Grid line spacing, Grid line color, Center on a Unit/Platform, Center on a location, Edit a location, View a location in the World map, and scroll with the scroll function.

Performance Steps

1. Select the Map button from the Function Bar or press the F1 key on the keyboard. The system will display the Map Control dialog box.

NOTE: The Map Button allows the user to change the background which gives the user more of less detail of the battlefield, add grid lines with different color and spacing, and center on a known point, platform or georeference. The Background Tab group is used to choose the Type, Scale, Zoom magnification, and appearance (brightness/Contrast) of the Map display. The Four background types are: (1) (CADRG) Compressed Arc Digital Raster Graphics, which is the Joint services standard map background product. (2) (VPF) Vector Product Format which is the map that displays major man-made and natural features on the SA screen such as cities, railroads, major roads, rivers, and lakes, (3) (DTED) Digital Terrain Elevation Data which is the uniform matrix of terrain elevation values, and provides basic quantitative data for all Military systems that require terrain elevation, slope, and surface roughness. (4) (Imagery) which is Reconnaissance Imagery that can be sent to the FBCB2 and displayed. The Map Control dialog box has standard push buttons at the bottom of each tab: Set Defaults, which sets the current settings as the Map default settings; Restore Defaults, which changes the current settings to the map default settings; the OK button, which applies the changes and closes the dialog box; the Apply button, which applies the changes and keeps the dialog box open; the Close button, which closes the dialog box; and the Help button, which provides help on the background tab.

2. Select the Background tab if it is not selected. The system will display the Background tab group.
   a. Deselect CADRG by unchecking the box to the left of CADRG.

NOTE: You must deselect one background type before you select another due to the fact that if you have more then on type selected, you might cover information from one map with the other.

   b. Select the VPF by placing your cursor in the box to the left of VPF and clicking, which will put a checkmark in the box.
   c. Select Apply. The system will display a light tan colored map with Cities, Lakes, Rivers, Railroads, Names, and Roads if they are on the map.
   d. Select the edit button to the right of the VPF. The system will display a VPF dialog box.

NOTE: From here you can turn off one or more of the attributes by deselecting whatever one the user don't want displayed.

   e. Deselect the names and select apply.

NOTE: Notice how the Names that were displayed on the map are no longer displayed. If the user wants it back on the map then all the user needs to do is select the box next to the attribute and select apply.

   f. Select close. The system will redisplay the Map Control dialog box.
   g. Deselect the VPF and select DTED.
   h. Select the apply button. The system will display a Dark green, Light green and Brown map.

STP 11-25C13-SM-TG

Performance Steps

NOTE: There are no edit functions for the CADRG, DTED, or Imagery.

    i. Deselect the DTED and select the Imagery and select apply. The system will display an aerial photograph of the earth surface as seen from above.
    j. Deselect the Imagery and select the CADRG and select apply. The system will display the standard map.
    k. Select the Scale pull-down button. The system will display the different scales that can be selected between 1 meter and 10 kilometers.
    l. Select 250K and select apply. The system will display a tannish colored map in a scale of 250K.

NOTE: Notice that it is hard to read. You can use the zoom button to enlarge the image.

    m. Select the Zoom pull-down button. The system will display the different zoom magnifications between 1/4X to 8X.
    n. Select the 2X and select the apply button. The system will display the map in 2X magnification.
    o. Select the Brightness slide bar and drag it left or right to adjust the brightness of the map display.
    p. Select the Contrast slide bar and drag it left or right to adjust the Contrast.
    q. Select the Zoom pull-down menu and select 50K and select apply.
    r. Select the Zoom pull-down button and select 1X and select apply. The system will display the standard map.

3. Select the Grid tab. The system will display the Grid tab group.

NOTE: The Grid tab allows the user to customize the map display grid lines. The Grid tab contains two areas: Grid type and Grid Lines. The Grid Type area includes two combo boxes: Coordinate Type and MGRS Accuracy.

    a. Select the Coordinate pull-down arrow. The system will display the coordinate types.

NOTE: The four Coordinate types are: Military Grid Reference System (MGRS), Longitude and Latitude (LAT/LON), Degrees, Minutes, Seconds (DMS), Universal Traverse Mercator (UTM).

    b. Select the MGRS if it is not already set to that type, which is the standard, coordinate type for the military. The system will display the MGRS in the Coordinate type text box.
    c. Select MGRS Accuracy pull-down arrow. The system will display 1, 10, and 100 meters.
    d. Select 1m. The system will display 1m in the MGRS Accuracy text box.
    e. Select the box to the left of Show Grid to check the box. The system will show a checkmark in the box and activate two additional pull down arrows: Grid line Spacing and Grid line color.

NOTE: Grid line Spacing provides the user options based on the Coordinate type selected. The Grid line Color gives the user the option of changing the color of the grid lines.

    f. Select the Grid line spacing pull-down arrow. The system will show the different selections for spacing.
    g. Select 1km which is the standard distance that the Military uses between grid squares
    h. Select the Grid line color pull-down arrow. The system will show the different colors for the grid lines.
    i. Select black which is the standard color that the Military uses for a Map.
    j. Select Apply. The system will display the grid lines in the color and distance that the user chose.
    k. Deselect the Show grid lines.
    l. Select Apply.

4. Select the Center tab. The system will display the Center tab group.

NOTE: The Center Tab will display three additional tabs: the Unit/Platform tab, Location tab, and the Scroll tab. The Center tab gives the user the ability to move quickly around the map. The Center function allows the user to move the map around to a specific unit or a general location.

Performance Steps

5. Select the Unit/Platform tab if it is not already selected. The system will display the Unit/Platform tab group.

NOTE: The Unit/Platform Tab allows the user to center the map on any units, platforms, or georeference that exist as SA within the visible battle space. The Unit/Platform tab contains: Refresh button, which refreshes or updates the unit platform options. The Search box allows the user to find the desired unit in a quick manner. The Details button allows the user to get information about a selected icon. The pull-down arrow allows the user the ability to center on Friendly, Observed, Air and Georeference one time unlike the Auto Center button which continually keeps the user's own platform centered. If no units of the type exist, the system does not display any units.

    a. Select the Unit/Platform pull down arrow just under the virtual keyboard. The system will display four selections.

NOTE: There must be some friendly, observed, Air and georeference on the map in order to see how the next steps work. The user can place some of these on the map by the SALT report in Combat Messages. See Task 171-147-0001 Step 2. After putting some icons on the map go back to the Map button, Center tab, Unit/Platform tab and select the refresh button.

    b. Make a selection from the list such as Observed. The system will display the observed/hostile/unknown units and platforms available to the user in the window under the Unit/Platform tab.
    c. Select one of the available Units or platforms from the window under the Unit/Platform tab. The system will highlight that selection.
    d. Select the Apply button. The system will center on the Unit/Platform that the user selected.
    e. Select the Unit/Platform pull down arrow. The system will display the same four selections.
    f. Select Friendly. The system will display Friendly in the text box and display all the friendly Units/Platforms that are on the map in the window below the Unit/Platform tab.
    g. Select one of the Unit/Platforms from the list. The system will highlight the selection.
    h. Select the Apply button. The system will center on that Unit/Platform.

6. Select the Location tab. The system will display the Location tab group with a list of all the available, pre-loaded, Map Data Groups and Georeference.

NOTE: The Location tab gives the user five methods of changing the map display area: Map data groups/Georeference which is the window with the list of all the available, pre-loaded, map data groups and georeference, Fill Location button which allows the user to pick a location using the mouse cursor, Manual Fill location text box which allow the user to type in the Location, The Edit Locations button which allows the user to create and modify a desired map data group, and World Coverage button which allows the user to choose and load a SA map anywhere in the world provided that the map area is loaded into the software.

    a. Select the Fill Loc button. The Map control dialog box will disappear and the mouse cursor will be replaced with a cross hair.
    b. Select a location on the map. The system will display the Map control dialog box with the location in the Location text box.
    c. Select apply. The map will center on the location that you picked.

NOTE: Move the Map Control dialog box to the side if it is in the way to see that it centered on the location that the user picked.

    d. Select the Edit Locations button. The system will display the Edit Map Locations dialog box.

NOTE: The Edit Locations button allows the user to make a map data group for ease of movement on any loaded map.

    e. Select the Group Name text box and type a name that the user wants to call the folder or select the virtual keyboard and type the name using the mouse and select OK.

Performance Steps

NOTE: The group name is limited to 20 characters.

    f. Select the Location Label text box and type the name that the user wants to call the location or select the virtual keyboard and use the mouse to type the name and select the OK button.

NOTE: The Location Label is limited to 20 characters.

    g. Select the Fill Loc button and select a location on the map. The Edit Map Locations dialog box will reappear with the location in the Location text box.

    h. Select the Apply button. The Edit Map Location dialog box will display the new group folder in the window with the Location name under the folder.

NOTE: Now all the user has to do is select the group from the Location tab and center on it.

    i. Select close. The system will display the Location Tab group.
    j. Select the World Coverage button. The system will display the World Coverage dialog box.

NOTE: The World Coverage allows the user to choose and load a SA map anywhere in the world provided the map area is loaded in the software. Currently all maps are not loaded. There are two tabs in the World Coverage dialog box: Background and Scroll. There are several common push buttons that appear in each of the two tabs. They do the same thing in both tabs. They consist of: Zoom Out, which zooms the map back out once you have zoomed in on a portion of the map, Previous View which lets the user go back one view at a time, Set Center which allows the user to quickly set center anywhere in the world, World View which quickly displays the world map, Close which closes the World Coverage dialog box, and Help which give the user help on the World Coverage dialog box.

    k. Select the Background tab which has two groups: Types and Scales.
    l. Select the type of map that you want to see (CADRG, DTED, IMAGERY) by checking the box next to the name.

NOTE: If the User wants to see the World map in the Imagery mode, the user must go to the Background tab in the Map Control dialog box and uncheck the map type that it is in and then check the Imagery type and select OK and then go back to the Center tab, Location tab and select the World Coverage button.

    m. Select the scale by checking the box next to the scale that you want. The system will display a checkmark in the box and a yellow blotch on the map where that scale applies.

NOTE: The user can only have one scale checked at once so the user must uncheck the scale that the user doesn't want.

    n. Select Set Center button and move your cursor over the world map. The cursor changes to a square box.
    o. Place the cursor over the area on the map that the user wants to center on and click the left mouse button. The system will center the SA map on that location.
    p. Select the Close button. The system will display the Map Control dialog box.
    q. Double-click on the Defaults folder on the Location Tab, Or select the + sign to the left of the Defaults folder to open the folder.

NOTE: Notice the list of Different areas that you can center on.

    r. Select a location from the list. The system will highlight the selection.
    s. Select the Apply button. The system will center the map on that area and leave the Map Control dialog box open.
    t. Select the Close button. This will close the Map Control dialog box.

  7. Select the Scroll Tab. The system will display the Scroll Tab group.

NOTE: This function allows the user to scroll the map display in the direction of the arrow one full screen at a time. This is used when the user needs to view an area close by.

STP 11-25C13-SM-TG

Performance Steps
    a. Select any arrow and watch the map move one screen in the direction that was selected.
    b. Select the Close button. The system will return to the Ops screen.

Performance Measures   <u>GO</u>   <u>NO GO</u>

1. Select the Map button from the Function Bar or press the F1 key on the keyboard.   —— ——

2. Select the Background tab if it is not selected.   —— ——

3. Select the Grid tab.   —— ——

4. Select the Center tab.   —— ——

5. Select the Unit/Platform tab if it is not already selected.   —— ——

6. Select the Location tab.   —— ——

7. Select the Scroll Tab.   —— ——

Evaluation Guidance: Score the Soldier GO if all steps are passed. Score the Soldier NO-GO if any step is failed. If the Soldier scores NO-GO, show him what was done wrong and how to do it correctly.

EMPLOY FIPR FUNCTIONS USING FBCB2 VERSION 3.4
171-147-0019

Conditions: Given a vehicle with an operational Force XXI Battle Command Brigade and Below (FBCB2) system with software version 3.4 loaded, precision lightweight global positioning system receiver (PLGR), enhanced position location radio system (EPLRS) if equipped, and single-channel ground air radio system (SINCGARS) with Internet controller (INC).

Standards: As a minimum you must: Employ screen operations functions to manage and exchange information, to include: Displaying messages in the FIPR queue; Identifying if the message is an Alert, the time the message was sent, the Message type, if an Operator response is required, who the Source Originator is, the Address type, type of Danger Zone, distance to the danger zone, direction to the danger zone, location of the danger zone, Originator of the danger zone, and view the alerts in the marquee.

Performance Steps

NOTE: Messages of each precedence must be received in the FIPR Queue before this task can be accomplished.

1. Select the FIPR (flash, immediate, priority, routine) button on the Function bar. The system will display the FIPR dialog box.

NOTES:

(1) The FIPR button allows the user to check messages that the system receives. An audible alarm of short duration alerts the user to incoming messages. The number at the end of the "FIPR" button represents the total number of messages in the queue. The system ranks messages by precedence as set by VMF (variable message format). A black exclamation mark (!) highlighted in yellow on the FIPR button informs the user that one or more warning messages are on the queue. A black plus (+) mark highlighted in yellow informs the user that an operator response is requested.

(2) There are six buttons on the FIPR Queue no matter what tab the user is in. The Display button, which allows the user to display the message, the Delete button, which allows the user to delete the message that is highlighted, the Delete All in Tab, which allows the user to delete all messages in the tab that is selected, the Refresh button, which allows the user to refresh the FIPR queue if a message arrived while viewing other messages, which is indicated by a black flag highlighted in yellow, the Close button which closes the FIPR Queue, and the Help button, which allows the user to get help on the FIPR function.

2. Select the FLASH tab if it is not selected already. The system will display the Flash tab group.

STP 11-25C13-SM-TG

Performance Steps

NOTES:

(1) The FIPR dialog box contains five tabs. Each tab represents message precedence in their order of rank. The number on each tab indicates the number of messages for that level of precedence. Flash is the highest, then Immediate, Priority, Routine, and Warnings, which contains Warnings/Alerts, Cautions, and Danger Zones. If no messages have been sent to the system, a 0 will be displayed on the FIPR button.

(2) Each tab in the FIPR queue (Flash, Immediate, Priority, Routine) has six message headers which include: (1) Alert which informs the user if the message is an Alert or not. (2) Time, which informs the user of the time that the message was sent. (3) Sec, which informs the user of the Security classification of the message. (4) Msg Type, which informs the user of the Type of message. (5) OR which informs the user that an operator's response is required. This column will display a Y when a response is required. (6) Source Originator, which informs the user who sent the message. The last tab, which is the Warnings tab, contains information on Warnings/Alerts, Cautions, and Danger Zones.

   a. Select the message from the list by left clicking it with the mouse pointer. The message will highlight.
   b. Select the Display button. The dialog box for the type of message that was selected will display.

NOTE: If a LOG Report was selected in sub-step a., then the LOG Report dialog box will display.

   c. Select the Close button for the message that was displayed. The system will return to the FIPR Queue on the same tab that the user was on.

3. Select the Immediate tab when you want to read an immediate (second highest precedence) message.

NOTE: The system displays the exact same dialog box except that you are in the Immediate tab now. All functions are the same for each of the tabs except for the Warnings Tab.

   a. Select the message from the list by left clicking it with the mouse pointer. The message will highlight.
   b. Select the Display button. The dialog box for the type of message that was selected will display.
   c. Select the Close button for the message that was displayed. The system will return to the FIPR Queue on the same tab that the user was on.

4. Select Priority tab when you want to read a priority (third highest precedence) message.
   a. Select the message from the list by left clicking it with the mouse pointer. The message will highlight.
   b. Select the Display button. The dialog box for the type of message that was selected will display.
   c. Select the Close button for the message that was displayed. The system will return to the FIPR Queue on the same tab that the user was on.

5. Select the Routine tab when you want to read a routine (lowest precedence) message.
   a. Select the message from the list by left clicking it with the mouse pointer. The message will highlight.
   b. Select the Display button. The dialog box for the type of message that was selected will display.
   c. Select the Close button for the message that was displayed. The system will return to the FIPR Queue on the same tab that the user was on.

6. Select the Warning tab. The system will display the Danger Zones and Marquee tabs.

Performance Steps

NOTE: The Warnings Tab contains a window that displays the warnings received by the system. Under the Warnings Tab, on the FIPR dialog box is a Danger Zones Tab. The FBCB2 system creates these zones when the operator receives certain messages. As the operator's vehicle approaches these danger zones, the FBCB2 system warns the operator with an audible tone and displays the alert on the Warnings/Alerts marquee. The Danger Zones tab has five columns. (1) Type, which contains the type of Danger Zone. (2) Dist, which contains the distance to the Danger zone from the Platforms location. (3) Dir, which contains the direction to the Danger zone. (4) Location, which contains the grid location of the Danger zone. (5) Originator, which contains the name of the message originator.

    a. Select the Danger Zones tab if it is not selected. The system will display the Danger Zones tab group.

NOTES:

(1) A Warning/Alert must first be sent to the users system that is close to it's platform's location before the next step can be performed such as a MOPP alert. Once the users platform gets that warning, it should show in the FIPR Queue and on the Warnings/Alert Marquee.

(2) All of the Tab groups in the FIPR dialog box have common push buttons at the bottom of the box. Display button shows the highlighted message. Delete button deletes the highlighted message, Delete all in tab button deletes all the messages in the tab, Refresh button updates the Queue when the system receives a new message while the user was viewing the messages by showing a black flag highlighted in yellow on the Refresh button, Close button closes the dialog box and Help button helps the user with the FIPR Queue button.

    b. Select the Danger Zones tab. The system will display the Danger Zones tab group.

NOTE: If there is no warnings such as an NBC1 report listed in the Danger Zones tab then go to the Messages Button and create a NBC1 and locate it close to your platform icon on the map then send it out. Once this happens it should show in the Danger Zones tab.

    c. Select one of the danger zone messages in the Danger Zones tab by highlighting it. The system will show the message highlighted in black.
    d. Select the Details button at the bottom of the Danger Zones tab. The system will display the Hook dialog box, which shows the content of the danger zone.

NOTE: From the Hook dialog box the user can delete, and edit the message.

    e. Select the Close button. The system will return to the ops screen.
    f. Reselect the FIPR button and then the Warnings tab. The system will display the Warnings tab group.
    g. Select the Marquee tab. The system will display the Marquee tab group.

NOTE: The Marquee tab allow the user to view all the warnings and alerts together on one page.

    h. Select the Close button. The system will return to the ops screen.

Performance Measures	GO	NO GO
1. Select the FIPR (flash, immediate, priority, routine) button on the Function bar.	——	——
2. Select the FLASH tab if it is not selected already.	——	——
3. Select the Immediate tab when you want to read an immediate (second highest precedence) message.	——	——
4. Select Priority tab when you want to read a priority (third highest precedence) message.	——	——

Performance Measures   <u>GO</u>   <u>NO GO</u>

   5. Select the Routine tab when you want to read a routine (lowest precedence) message.   —— ——

   6. Select the Warning tab.   —— ——

Evaluation Guidance: Score the Soldier GO if all steps are passed. Score the Soldier NO-GO if any step is failed. If the Soldier scores NO-GO, show him what was done wrong and how to do it correctly.

STP 11-25C13-SM-TG

EMPLOY STATUS FUNCTIONS USING FBCB2 VERSION 3.4
171-147-0020

Conditions: Given a vehicle with an operational Force XXI Battle Command Brigade and Below (FBCB2) system with software version 3.4 loaded, precision lightweight global positioning system receiver (PLGR), enhanced position location radio system (EPLRS) if equipped, and single-channel ground air radio system (SINCGARS) with Internet controller (INC).

Standards: As a minimum you must: Employ screen operations functions to manage and exchange information, to include: Determine if a component is degraded, determine the status of the network, Situational awareness data, the count of Observed and Friendly icons, and the status of the Hard drive.

Performance Steps

1. Select the Status button from the Function Bar or the F5 button on the Keyboard. The system will display the Status tab group.

NOTE: The Status function provides a means for the user to observe the operational status of the Platform's communications systems, the Tactical Internet (TI), and the general disk employment. The Status tool should also be used as a quick reference and troubleshooting tool when it is noticed that the system is not operating at its optimal level. The Status Dialog box contains three Sub-tabs: the Systems, SA, and General tab. The Status button contains read-only information.

2. Select the Systems tab. The systems tab displays two columns of information: Systems Name and Status.

NOTE: System Name has small file folders that represent the user platform's Global Positioning System (GPS), Local Communications (LOCAL COMM), and other associated systems connected to the platform, to include the Laser Range Finder (if applicable). The second tab is the Status tab which displays the diagnostic details for the associated component, which are: Go (Green) diagnostic test indicated that the component is operative at the optimal level, Degraded (Amber) diagnostic test indicated that the test is operative at an acceptable level, No GO (RED) diagnostic test indicated that the component is inoperative or not operative within acceptable parameters, Not Tested (WHITE) device unavailable or not configured for the system.

    a. Select the + sign to the immediate left of the GPS folder symbol, Or double-click on the folder symbol itself to open the contents. You should see the associated sub-components and their corresponding status indications. If there is no + sign showing then there is no information available for that component.

NOTE: You should see the following status indications: TIME- displays the Time Figure of Merit (TFOM) quality, HEADING- displays the quality of the heading received from the GPS, POSITION- displays the Figure of Merit (FOM) quality

    b. Close the GPS Component by selecting the - sign to the immediate left of the GPS folder symbol, Or double-click the GPS folder symbol.

    c. Open the LOCAL COMM folder by selecting the + sign, or double-clicking the LOCAL COMM folder.

STP 11-25C13-SM-TG

Performance Steps

NOTE: The Local Comm status tells the user how the communication devices located on the platform are functioning. The List of devices and interfaces will vary based on the platform role and configuration. Opening the LOCAL COMM displays the sub files with the diagnostic results for that system. The sub files reveal the names of the pertinent devices. The devices you should see are: Local Area Network (LAN) used to connect local FBCB2 systems by way of a continuous cable, REMOTE DISPLAY which is a remote that can be dismounted. Router_1, which is a special-purpose computer (or software package) that transfers data between two or more networks that use the same protocols. Routers look at the destination addresses of the information packets and route them to the proper FBCB2 system. The FBCB2 uses the INC to perform the duties of the router.

    d. Select the + sign next to the Router 1 folder, or double-click the Router 1 folder to open it.

NOTE: Router_1, which is a special-purpose computer (or software package) that transfers data between two or more networks that use the same protocols. Routers look at the destination addresses of the information packets and route them to the proper FBCB2 system. The FBCB2 uses the INC to perform the duties of the router. Router_1_PPP is Point-to-Point Protocol. It allows a computer to use TCP/IP (Internet) protocols. A protocol is a definition of how computers will act when talking to each other.

    e. Open the R1_SINCGARS1 folder by selecting the + sign or double-clicking the folder.

NOTE: R1_SINCGARS1 and R1_SINCGARS2 are the radios that are connected to the FBCB2 system. The sub-components are Interface which is an expansion board within the radio system that permits connection of the external devices, the Net_ID/Frequency which is the Network identification numbers and or frequency, the Pckt_Mode which enables the transmission and reception of data blocks containing control information- such as routing, address, and error control, as well as data.

    f. Open the R1 EPLRS_Radio by selecting the + sign or double-clicking the folder.

NOTE: The R1 EPLRS_Radio is the Enhanced Position Location Reporting System which is an integrated Command, Control, and Communications (C3) system that provides near real-time data communications, position/location, navigation, identification and reporting information on the modern battlefield. The sub-components are R1 EPLRS_LCN_1, LCN_2, and so on depending on how many channels. The Logical Channel Number (LCN) packet switched networks allocate a number at the time a call is set up, which distinguishes packets belonging to one call on a link from all others.

    3. Select the SA Tab. The system will display the SA tab group.

NOTE: The SA Tab function is used to give the user information on the Current SA server, the Broadcast net, SA net member count, switching of nets 1, 2, and CSMA, Server connectivity and status on TI location quality.

    a. Select the Net 2 radio button. The system will switch to that net if your platform has that capability.

NOTE: When you switch the net, all the information on the SA tab also switches.

    b. Select the Net 1 radio button to switch back.

    4. Select the General Tab. The system will display the status of the Disk Utilization.

NOTE: The General tab displays a pie chart of the system's disk utilization in percentages and total disk capacity. Percentages of disk space used and free indications appear to the right of the pie chart.

    a. Select the refresh button to refresh the Information.
    b. Select the Close button to close the dialog box.

Performance Measures   <u>GO</u>   <u>NO GO</u>

1. Select the Status button from the Function Bar or the F5 button on the Keyboard.   ——   ——
2. Select the Systems tab.   ——   ——
3. Select the SA Tab.   ——   ——
4. Select the General Tab.   ——   ——

Evaluation Guidance: Score the Soldier GO if all steps are passed. Score the Soldier NO-GO if any step is failed. If the Soldier scores NO-GO, show him what was done wrong and how to do it correctly.

STP 11-25C13-SM-TG

EMPLOY ADMIN FUNCTIONS USING FBCB2 VERSION 3.4
171-147-0021

Conditions: Given a vehicle with an operational Force XXI Battle Command Brigade and Below (FBCB2) system with software version 3.4 loaded, precision lightweight global positioning system receiver (PLGR), enhanced position location radio system (EPLRS) if equipped, and single-channel ground air radio system (SINCGARS) with internet controller (INC).

Standards: As a minimum you must: Employ screen operations functions to manage and exchange information, to include: setting the Platform location and Quality manually, setting the Course, Speed, Elevation and Altitude if applicable manually, Determine if the proper Data Net Frequency and Radio Set ID is set, setting the Requesters call sign and Medevac Voice Net Frequency, setting the Display/Message and audio settings, Local time zone, Time and Motion filters, and the Reporting mode.

Performance Steps

1. Select the Admin Button (F6). The system will display the Admin dialog box.

NOTE: The Exit Ops, Destroy FBCB2, OK, Apply, Close, and Help buttons at the bottom of the Admin dialog box stay displayed when cycling through each of the tabs in the Admin dialog box. Those buttons do the same thing in each tab.

(1) The Exit Ops button allows the user to exit the Ops screen and go back to the session manager screen.

(2) The Destroy FBCB2 button allows the user to destroy the FBCB2 by overwriting the files.

(3) The Ok button allows the user to make the changes and closes the dialog box.

(4) The Apply button allows the user to make the changes and keeps the dialog box open.

(5) The Help button allows the user to get help on the Admin function.

2. Select the Platform Settings Tab. The system will display two additional tabs, the Location and the Misc tab.

NOTE: This function allows the user to manually enter the vehicle/platform location. The PLGR usually provides this information. If the PLGR is inoperable, the user can manually enter the required settings. If the platform is fitted with an EPLRS, the Net Control Station reports the EPLRS location and compares the quality of SA on the Tactical Internet.

3. Select the Location tab. The system will display the Location tab group.
   a. Select the Location text box and type in the grid where you want your platform to be, or select the Fill Loc Button and select a position on the Map. The system will display your icon on the map.

NOTE: If the Admin dialog box is in the way of your icon, left click the dark gray area at the top of the Admin dialog box and drag it to the side.

   b. Select the Quality Down arrow and make a selection from the list.

NOTE: If a PLGR was connected to the FBCB2 and functioning correctly, this information would be entered automatically. The Quality is the degree of accuracy, which is the Figure of Merit on the PLGR.

   c. Select the Course text box and type in a direction in degrees or select the virtual keypad and type the degrees in using the mouse pointer select any degree between 0 and 359 and select OK. The system will return to the Admin dialog box.
   d. Select the Speed text box and type in the speed in Kph between 0 and 2047 or use the virtual keypad as before.

Performance Steps
  e. Select the Elevation text box and type in the elevation in feet between 1 and 65536 or use the virtual keypad.
  f. Select the Apply button. The system will apply the entries.

NOTE: Notice how your Platform icon is displayed on the map.

 4. Select the Misc tab. The system will display the Misc tab group.

NOTE: If the Requester's Call sign and MEDEVAC Voice Net Frequency are not entered before attempting to utilize the MEDEVAC Combat message, the user will be unable to send a MEDEVAC Combat massage. The Data Net Frequency field is a view only field containing the programmed SINCGARS frequency for the Tactical Internet. This Frequency will vary from unit to unit. The Radio Set ID is a view only field containing the programmed EPLRS radio frequency.

  a. Select the Requester's Call Sign text box and type in the Call sign, Or select the virtual keyboard and type the call sign in.

NOTE: When typing the call sign, it must be at least 17 characters long or it will not accept it. If your call sign is only 3 characters then type those 3 characters and hit the space bar 14 more times.

  b. Select the MEDEVAC Voice Net Frequency text box and type the frequency, Or select the virtual keyboard and type the frequency and select OK.

NOTE: When typing the MEDEVAC Voice Net Frequency, it must be at least 8 characters long or it will not accept it. If the frequency is only 5 characters long, then type the 5 characters and select the space bar 3 more times.

  c. Select the apply button.

NOTE: If the entries were entered correctly it would apply them. If they were entered incorrectly an error message would appear explaining what the problem is.

 5. Select the Local Settings tab. The system will display Two additional tabs, the Display/Message, and the Audio tabs.
  a. Select the Display/Message Tab. The system will display the Display/Message tab group.
  b. Select the Chembio Auto Send radio button On/Off.

NOTE: This function will be grayed out if your vehicle is not equipped with the (MICAD) Multipurpose Integrated Chemical Agent Detector or the (LRBSDS) Long Range Biological Stand-Off Detection System.

  c. Select the Reminder Dialog radio On button. The On button will highlight.

NOTE: The Reminder Dialog function toggles all Periodic Reminders and Voice on/off.

  d. Select the Warning Time Interval pull down arrow. Make a selection from the list between 3 seconds and 10 seconds.

NOTE: This is the time that each Alert/Warning message will be displayed in the Alert/Warning Marquee on the Classification/Status Bar.

  e. Select the Local Time Zone pull down arrow and scroll to your time zone.
  f. Select the Apply button to make the changes.

 6. Select the Audio tab. The system will display the Audio tab group.

NOTE: The Audio button allows the user to set Alerts, Notices, and Reminders to Tone, Voice, or Off. It also sets the voice volume and to mute alerts or preview a voice alert.

  a. Select the Tone, Voice, or Off radio button under Alerts depending on if you want to hear a tone, voice or nothing on the speaker.
  b. Select the Tone or Off radio button under Notices if you want to hear Notices.

STP 11-25C13-SM-TG

Performance Steps

    c. Select the Tone1, Tone2, or Off radio button under the Reminders.

NOTE: There are two different tones that you can hear for Reminders.

    d. Select the Voice volume slider by left clicking and holding. Move the slider right to increase and left to decrease the volume.

    e. Check the Mute all box if you don't want to hear anything.

NOTE: Mute All is not recommended.

    f. Select the Preview Voice pull-down arrow. Make a selection from the list.

    g. Select the Play button and listen to the speaker.

NOTE: If you want to hear a voice, make sure that Alerts is set to voice.

    h. Select the Apply button to make the changes.

7. Select the SA Settings Tab. The system will display the SA Tab Group.

8. Select the Own tab. The System will display the Own Tab Group.

NOTE: This feature allows the user to set their own Time and Motion filters and the reporting Mode settings.

    a. Select the Time Filter pull down arrow and make a selection from the list between 10 seconds and 60 minutes.

    b. Select the Motion Filter pull down arrow and make a selection from the list between 50 meters and 2500 meters.

    c. Select the Reporting Mode radio button: Auto, Manual, or Off.

    d. Select the Apply button.

9. Select the Friendly Tab. The system will display the Friendly Tab Group.

    a. Select the Stale pull down arrow and make a selection from the list between 5 minutes and 2 hours. The system will display the time in the Stale text box.

    b. Select the Old pull down arrow and make a selection from the list between 10 minutes and 4 hours.

NOTE: The Old feature allows the user to set the elapse time before a friendly SA symbol becomes old (grays out on the map). If the system does not update its position report in the prescribed time, the system marks the last position as old.

    c. Select the Purge pull down arrow and make a selection from the list between 1 hour and 20 hours. The system will display the time in the Purge text box.

NOTES:

(1) The Purge Time is the elapse time before your system purges the symbol from your map (deletes it off your map) if the friendly system doesn't send its position report.

(2) You must select the settings according to your SOP.

(3) Restore all SA Default settings allows the user to restore the settings back to the default

    d. Select the Apply button.

10. Select the Observed Tab. The system displays the Observed Tab group. Complete the Observed tab exactly like the Friendly tab.

11. Select the Air tab. The system will display the Air tab group. Complete the Air tab exactly like the Observed tab.

    a. Select the Apply button. The system will save the settings.

    b. Select the OK button. The system will return to the Ops screen.

Performance Measures	GO	NO GO
1. Select the Admin Button (F6).	——	——
2. Select the Platform Settings Tab.	——	——
3. Select the Location tab.	——	——
4. Select the Misc tab.	——	——
5. Select the Local Settings tab.	——	——
6. Select the Audio tab.	——	——
7. Select the SA Settings Tab.	——	——
8. Select the Own tab.	——	——
9. Select the Friendly Tab.	——	——
10. Select the Observed Tab.	——	——
11. Select the Air tab.	——	——

Evaluation Guidance: Score the Soldier GO if all steps are passed. Score the Soldier NO-GO if any step is failed. If the Soldier scores NO-GO, show him what was done wrong and how to do it correctly.

STP 11-25C13-SM-TG

EMPLOY APPS FUNCTIONS USING FBCB2 VERSION 3.4
171-147-0022

Conditions: Given a vehicle with an operational Force XXI Battle Command Brigade and Below (FBCB2) system with software version 3.4 loaded, precision lightweight global positioning system receiver (PLGR), enhanced position location radio system (EPLRS) if equipped, and single-channel ground air radio system (SINCGARS) with Internet controller (INC).

Standards: As a minimum you must: Employ screen operations functions to manage and exchange information, to include: Driver's Display, Line of Sight, Periodic Reminders, Personnel Status Report, and Radio Net Join.

Performance Steps

1. Perform Driver's Display Function.
   a. Select the Apps button from the Function Bar or select the F7 button from the keyboard. The system will display the Apps dialog box.

NOTE: The Apps function provides the user with many unique applications and tools. The Apps dialog box has Two sub-tabs: the FBCB2 tab and the Misc tab which is not available.

   b. Select the FBCB2 tab if it is not already selected. The system will display the FBCB2 Applications list.
   c. Select Driver's Display and select the execute button. The system will display the Driver's Display: Compass dialog box.

NOTE: The Driver's Display gives the user the ability to turn the compass on for the driver to follow. The dialog box has two push buttons: The Navigation button, which allows the user to switch to the Navigation Dialog Box, and the Close button, which closes the dialog box. The Nav button does the same thing as the Navigation function learned earlier in the task.

   d. Select the Close button. The system will return to the Ops screen.
   e. Select the Apps button on the Function bar or select the F7 button on the keyboard. The system will display the Apps dialog box.
   f. Select the Close button. The system will return to the Ops screen.

2. Perform the Line of Sight Function.

NOTE: The Line of Sight function allows the user to draw a line on the SA display map from the Start (S) point to the End (E) point and view a graphic representation of the topography between the two points. The Function bar has several push buttons consisting of: Coordinate button which is a toggle button that shows the different types of coordinates, OK button which is used once you have selected your start and end points, Cancel to cancel the last action, Delete to delete the Start and end point, Help to get help on the Line of Sight tool, Virtual keyboard to type the coordinates in the Location text box, Add which is used to add point if you are using the type mode to enter coordinates, Named button which is for selecting pre-loaded locations.

   a. Select the Apps button from the Function Bar or select the F7 key from the keyboard. The system will display the Apps dialog box.
   b. Select the Line Of Sight function and select the Execute button. The system will display the Select 2 Points function bar on the right side of the map and the cursor as a cross hair.
   c. Select a start point on the map by left clicking a spot on the map. The system will display a white circle on that point.
   d. Select a second point on the map. The system will display a blue line connecting both of the points.
   e. Select OK. The system will remove the function bar and display the Line Of Sight dialog box.

Performance Steps

NOTE: The LOS dialog box displays the range, bearing, Start location, and end location of the line that you selected. It also has three push buttons consisting of Close to close the dialog box, Profile (SHOW) that brings up the Profile dialog box, and Help.

    f. Select the Profile (SHOW) button. The system will display the LOS Profile dialog box on top of the Line Of Sight dialog box.

NOTE: If the LOS dialog box is on top of the Line Of Sight dialog box, just drag the LOS dialog box over so both dialog boxes can be seen. The dialog box displays the Line of sight in the form of a graph from the start to the end point. Anything above the yellow line is dead space. It shows the distance in (Km) along the bottom and the Height in (Ft) on the left side of the dialog box.

    g. Select the Start point on the map by placing the cursor over the White circle with an S in it and left clicking the mouse, holding it and dragging the Start point around.

NOTE: Notice how the view changes on the LOS dialog box as the Start point is moved.

    h. Select the Close button. The system will return to the Ops screen.

3. Perform the Periodic Reminders function.

NOTE: The "Periodic Reminders" tool allows the user to create and store message reminders that will, at a users determined date and time, display a reminder dialog box with a user created text message. d time. There is an option to trigger an audio alert when the reminder is displayed. Zulu time (Greenwich Mean Time (GMT)) is the standard.

    a. Select the Apps button from the Function bar or the F7 key on the keyboard. The system will display the Apps dialog box with the FBCB2 tab selected.

NOTE: The Reminders dialog box has six common push buttons that are in each of the five tabs consisting of: OK which applies the changes and closes the dialog box, Apply which applies the changes and keeps the dialog box open, Defaults which sets the settings back to the original settings, Close which closes the dialog box, List which brings up a list of all reminders that have been set, and Help which gives the user help on the tab that is displayed.

    b. Select the Periodic Reminders by highlighting the name.
    c. Select the Execute button. The system will display the Periodic Reminders dialog box.

NOTE: The Periodic Reminders dialog box contains the following tabs, which is based on the frequency of the reminder: Once, which is used to create a one-time reminder. Daily which is used to create a reminder for each day, Weekly which is used to create a reminder for once a week, Monthly which is used to create a reminder for once a month, and Floating which is used to create a reminder for a particular time, day, and week among a range of months.

    d. Select the Once tab. The system will display the Once tab fields.
    e. Select the Reminder month radio button for the month that you want the reminder.

NOTE: The system defaults to the current month.

    f. Select the Reminder Time Hour slide button and hold the left mouse button and drag left to decrease the hour and right to increase the hour.
    g. Select the Reminder Time Minute slide button and hold the left mouse button and drag left to decrease the minute and right to increase the minute.
    h. Select the Reminder Day slide button and hold the left mouse button and drag left to decrease the day and right to increase the day.
    i. Select the Reminder Audio Yes/No radio button.

NOTE: Select Yes if you want to hear an audible tone when you are reminded or No if you don't want to hear a tone.

    j. Place your mouse pointer inside the Reminder Text box and click the left mouse button and type the text that you want to be reminded of.

STP 11-25C13-SM-TG

Performance Steps

    k. Select the Daily tab. The system will display the Daily tab fields.

NOTE: The Daily tab is done just like the Once tab. The only difference is there is no Month and Day field.

    l. Select the Weekly tab. The system displays the Weekly tab fields.

NOTE: The Weekly tab is done just like the Once tab. The only difference is there is you can select more than one day.

    m. Select the Monthly tab. the system will display the Monthly tab fields.

NOTE: The Monthly tab is done just like the Once tab. The only difference is that you can select more than one month to have the reminder.

    n. Select the Floating tab. The system will display the Floating tab fields.

NOTE: The Floating tab is done just like the Once tab. The only difference is you can select more than one month and you can select which week of the month to get the reminder.

    o. Select the Close button. The system closes the Create Reminders dialog box and displays the Ops screen.

4. Perform The Personnel Status Report.

NOTE: This function allows the user to update the Personnel Status Report and send it to higher Headquarters. At Battalion S1, the Personnel Status Report interfaces with the Combat Service Support Control System (CSSCS), which transmits personnel reports to higher echelons.

    a. Select the Apps button from the Function Bar or the F7 button from the keyboard. The system will display the Apps dialog box.
    b. Select the Personnel Status Report from the list with the mouse pointer. The system will highlight the Personnel Status Report.
    c. Select the Execute button. The system will display the Personnel Status Report dialog box.

NOTE: The Personnel Status Report dialog box has eight common push buttons which consist of: Message Addressing, which allows the user to set how and who they will send the message to, New which brings up a new Add personnel Record, Delete which deletes a record, Refresh, which refreshes the Data Display Area, Send which sends the report, Modify which allows the user to edit a record, Close which closes the Personnel Status Report, and Help which gives the user help on Personnel Status Reports. There is also a SEARCH BY pull-down arrow which allows the user to search for a record by: Last name, SSN number, Status, Unit name, Role ID, Grade, MOS, Gender, Nationality, Religion, Blood type and Effective DTG. There is also a search text box that allows the user to search for a record by typing the information in the text box and selecting the search button.

    d. Select the Message Addressing button. The system will display the Message Addressing dialog box.
    e. Enter the Message settings and Addresses as in task 171-147-0005, Apply Message Addressing features in FBCB2 Version 3.4
    f. After setting the Message Addressing settings, select the New button on the Personnel Status Report dialog box. The system will display the Add Personnel Record dialog box.

NOTE: The New button allows the user to enter information on new personnel by using the Virtual keyboard, Computer keyboard, Drop down arrows, and Radio buttons. Notice how the OK and Apply buttons at the bottom of the Add Personnel Record dialog box are grayed out. All the fields must be filled out before the buttons will become available.

    g. Select the Last Name text box with the mouse pointer or select the virtual keyboard and enter the last name in the text box.

Performance Steps

NOTE: If you select the text box, you can enter the information with the computer keyboard. If you select the virtual keyboard you must type the information by selecting the characters with the mouse pointer. The Last Name must not be more than 20 characters long.

    h. Select the First Name text box and enter the name.

NOTE: The First Name cannot be more than ten characters long.

    i. Select the Middle Name text box and enter the name if applicable.

NOTE: The Middle name cannot be more than ten characters long.

    j. Select the Suffix text box and enter the suffix if applicable.

NOTE: The Suffix cannot be more than three characters long.

    k. Select the SSN text box and enter the Social Security Number.

NOTE: You do not have to enter the dashes in the Social Security Number. The numbers will automatically move to the proper digit.

    l. Select the Nationality pull-down arrow and select the Nationality of the person that the user is entering the information for. The Nationality will display in the text box.
    m. Select the Religion pull-down arrow and select the Religion of the person.
    n. Select the proper radio button, either Male or Female for the person you entering information on. The radio button will highlight.
    o. Select the Blood Type pull-down arrow and select the Blood type for the person you entering information about.

NOTE: You must select the Whole Blood type, not the Platelet type.

    p. Select the Unit Name pull-down arrow and select the Unit that the person is in by highlighting the unit or by typing in the unit in the text box and selecting the search button.

NOTE: You must enter the Unit name exactly as it is displayed on the computer or it will not find the unit in the list.

    q. Select the Role/ID pull-down arrow and select the Role/ID from the list.
    r. Select the Grade pull-down arrow and select the grade of the person.
    s. Select the MOS pull-down arrow and select the MOS of the person.
    t. Select the Status pull-down arrow and select what the persons status is from the list.
    u. Select the Apply button to apply the new record. The system will display a new Add Personnel Record dialog box, Or Select OK, which will complete the record and return to the Personnel Status Report dialog box.

NOTE: Continue to do this until you have a record for all the personnel.

    v. Modify a Personnel Status Report.
    w. Select a record from the list on the Personnel Status Report dialog box. The record will highlight.
    x. Select the Modify button. The system will display the Modify Personnel Record(s) dialog box.
    y. Select the field that you want to edit or modify and make the changes.
    z. Select the OK button. The system will make the changes and display the Personnel Status Report dialog box.
    aa. Select the Send button to send the Report to the address that you entered in the Message addressing settings. The system will display a Personnel Report Sent dialog box stating that "THE Personnel Report was Sent."
    ab. Select the OK button. The system will return to the Personnel Status Report dialog box.
    ac. Select the Close button. The system will display the Ops screen.

5. Perform the Radio Net Join function.

STP 11-25C13-SM-TG

Performance Steps

NOTE: DO NOT select a new Network unless you intend to switch Networks because this function will reboot the computer and assume the new Network. The Radio Net Join function allows the user to re configure the communication subsystems on their own platform. Radio Net Join will automatically re configure the FBCB2 database, INC router, SINCGARS-ASIP and EPLRS. This is used when the users platform travels outside its SINCGARS or EPLRS network and must join a different network in order to continue to receive SA and C2 data. There are two areas in the Radio Net Join function: the SINCGARS Hop set and the EPLRS Configuration. The SINCGARS Hop set displays the current description and frequency identification. The EPLRS Configuration changes the network to a different units configuration.

    a. Select the Apps button from the Function Bar or the F7 key on the keyboard. The system will display the Apps dialog box.
    b. Select the Radio Net Join function from the list. The Radio Net Join will highlight.
    c. Select the Execute button. The system will display the Radio Net Join dialog box.
    d. Select the New FREQ_ID pull-down arrow under the SINCGARS Hop set. The system will display the SINCGARS Agent Table dialog box with the available Units and Net_ID frequencies.
    e. Select the Network that you want.
    f. Select the Change pull-down arrow under the EPLRS Configuration. The system will display the Unit Name dialog box.

NOTE: Here the user can select a new EPLRS network for FBCB2 communication purposes.

    g. Select the Network that the user wants to change to.
    h. Select OK if you actually want to switch networks or Close to stay in the same configuration.

Performance Measures	**GO**	**NO GO**
1. Perform Drivers Display Function.	___	___
2. Perform the Line of Sight Function.	___	___
3. Perform the Periodic Reminders function.	___	___
4. Perform the Personnel Status Report.	___	___
5. Perform the Radio Net Join function.	___	___

Evaluation Guidance: Score the Soldier GO if all steps are passed. Score the Soldier NO-GO if any step is failed. If the Soldier scores NO-GO, show him what was done wrong and how to do it correctly.

STP 11-25C13-SM-TG

EMPLOY NAV FUNCTIONS USING FBCB2 VERSION 3.4
171-147-0023

Conditions: Given a vehicle with an operational Force XXI Battle Command Brigade and Below (FBCB2) system with software version 3.4 loaded, precision lightweight global positioning system receiver (PLGR), enhanced position location radio system (EPLRS) if equipped, and single-channel ground air radio system (SINCGARS) with internet controller (INC).

Standards: As a minimum you must: Employ screen operations functions to manage and exchange information, to include: Creating a route, reversing the way points, setting Route Attributes and Analyze Route settings, Editing a route, Center on a route, Apply the Roll over Mode, turn on the Driver's display, and Select a Single point to navigate to.

Performance Steps

1. Select the NAV button. The system will display the Navigation dialog box.

NOTE: The Navigation function allows the user to create a new Route of March (ROM), select and edit a previously created ROM, reverse the direction of march on a selected route, set route attributes, analyze a route, center the SA map on the selected ROM, and transmit the Land Route Report. The "Roll over Mode" option when used with the driver's display will display the bearing and distance to the next way point as each way point is passed.

2. Select the Route Tab if it is not already selected. The system will display the Route Tab group.

NOTE: The Route ID function allows the user to select and display a route that has been previously created and saved by selecting the drop down arrow and highlighting the saved route. Once a saved route is selected, all of the other function buttons on the "Route" dialog box become active.

   a. Select the create button. The system will display the Create New Route dialog box.

NOTE: The Create New Route dialog box has three buttons: Map, List, and Cancel. The Map button allows the user to select locations on the map. The list button displays the Create Way points dialog box and allows the user to enter waypoints by typing them in the Enter Way point combo box.

   b. Place the cursor in the Enter New Route Name text box and select the back space key to delete the name that appears then type the name that you want to call the route up to 9 characters.

NOTE: You can also type the name by selecting the Virtual Keyboard and using the mouse to select the keys on the Virtual Keyboard or your fingers on the touch screen.

   c. Select the Map button. The system will display the Map with the Create Route Function Bar to the right and the cursor becomes a cross hair.
   d. Create the route by clicking a spot on the map where you want the route to start, and then continue to click along the route creating waypoints each time the mouse is clicked until you have plotted the route.

NOTES:

(1) Notice on the map that the points where you selected when creating the route have checkpoint symbols with the first one highlighted in white. You can move to the next waypoint by selecting the right arrow in the Navigation Dialog box.

(2) If the Navigation dialog box is in front of the route that you created, you can move the box by placing the cursor in the dark gray area of the Navigation dialog box and left clicking and holding the button and dragging the box to the side.

Performance Steps

  e. Select the OK button. The system will display the Navigation dialog box with the new name in the Route ID text box and all the buttons in the Navigation dialog box are active now.

 3. Select the Manage button. The system will display the Edit Route dialog box.

NOTE: The Manage button allows the user to rename, copy, or delete a route and modify an existing route. The user can rename a route by typing the name in the Name text box and selecting the Rename button. The user can delete a route by typing the name of the route that you want to delete in the Name text box and selecting the Delete button. The user can copy a route with a new name by typing the new name of the route in the Name text box and selecting the Copy button.

  a. Select the Map button. The system will display the Edit Route function bar with the route highlighted in blue with a white circle on waypoint one.

NOTE: If the user wanted to modify a different route then the one they are currently on then they would have to go back to the Navigation dialog box and select the route from the pull down arrow then select the Manage button.

  b. Place the cursor on a spot on the map after the last way point and left click the mouse. The system will add a waypoint to the existing route.
  c. Select the Delete button. The system will delete the last waypoint.

NOTE: Every time the user clicks on the map a new waypoint is added. If the user makes a mistake, they can select the delete button and delete the last waypoint.

  d. Select the OK button when finished with the modification. The system will display the Edit Route dialog box.

 4. Select the List button. The system will display the Edit waypoints dialog box with all the way points in the Way point List text box that you created earlier.

NOTE: The List button allows the user to delete a single waypoint, delete all waypoints from the list, or add additional waypoints to the list. Notice that some of the buttons are not active until a waypoint is selected for deletion or a waypoint is entered in the Enter Way point text box.

  a. Select the last waypoint in the list. The waypoint will be highlighted.
  b. Select the Delete button. The waypoint will be deleted from the list but not from the route on the map.
  c. Select the Save button. The system will delete the waypoint from the route on the map and return to the Edit Route dialog box.
  d. Select the List button. The system will display the Edit Way point dialog box.
  e. Place the cursor in the Enter Way point text box and left click the mouse.
  f. Enter a grid close to the grid of the last waypoint in the list.

NOTE: Ensure that you type the grid just like the waypoints that are in the list with the Grid Zone designator.

  g. Select the Insert button. The system will add the waypoint to the list but not to the route.
  h. Select the Save button. The system will add the waypoint to the Route and return to the Edit route dialog box.

STP 11-25C13-SM-TG

Performance Steps
NOTES:

(1) If the user wants to delete all the way points, the user would highlight one of the way points in the list and select the Delete All button and select the Yes button on the Confirm Deletion dialog box.

(2) The user can add way points by selecting the pull down arrow on the Edit Way points dialog box and select either Keyboard (Kbd), Laser Range Finder (LRF), Map, Own, Name, and Virtual Keyboard (Vkb).

    i. Select the close button from the Edit Route dialog box. The system will display the Navigation dialog box.

  5. Select the Reverse button several times.

NOTE: Notice how the Start Point on the route becomes the new Release point and the Release point becomes the new Start point.

  6. Select the Route Attributes button. The system will display the Route Attributes dialog box.

NOTE: The Route Attributes allows the user to describe the aspects of the route.

    a. Select the Route Description pull down arrow. The system will display a list.
    b. Select an option from the list. The option will be displayed in the text box.

NOTE: Do the same for the Route Classification, Movement Rate, Open/Concealed Indicator, and Route type.

    c. Select the Virtual Keypad to the right of the Overhead Clearance/Height (m) text box. The system will display the Virtual Keypad on the screen.
    d. Enter the Height in meters by selecting the appropriate keys and selecting the OK button. The system will display the Height in the text box.

NOTES:

(1) Notice in the Virtual Keypad at the bottom that there is a number that the user must stay within when entering a number.

(2) Do the same for Traveled Way Width (m), and Weight Classification (tons).

    e. Place the cursor in the Comments text box and left click the mouse.
    f. Type any comments that pertain to the Route Attributes if applicable.
    g. Select the OK button. The system will display the Navigation dialog box.

  7. Select the Analyze Route button. The system will display the Analyze Route dialog box.

NOTES:

(1) Analyze Route button allows the user to analyze an existing Route of March to determine the Flank Line of Sight, areas that exceed the vehicles maximum uphill and downhill degree of grade, degree of side slope for each segment of the route, the 360 degree Line of Sight at any point along the route, the travel time of each leg of the route and show a graphical profile of each segment of the route.

(2) The Route Data on the right side of the Analyze Route dialog box shows the name of the route, how many waypoints are on the route, and the length of the route in Km. The Segment Data shows the grid of the Start point and End point and the Segment length. To find out what the length of the other segments just click on the right arrow underneath the Show Segment Profile. Each time the arrow is selected, it will show the next segment.

STP 11-25C13-SM-TG

Performance Steps
    a. Select the Virtual Keypad under the Flank Line of Sight Data for the Range. The system will display the Virtual Keypad on the screen.
    b. Enter the appropriate range in meters between 1 and 100000 by selecting the numbers with the mouse and selecting the OK button. The system will display the numbers in the text box.

NOTE: Do the same for the Distance Above Ground (m), Maximum Uphill Grade (%), Maximum Downhill Grade (%), Max Slope (%), and Rte Width (m).

    c. Select the Color pull down arrow and select a color from the list. The color will be displayed in the Color text box.

NOTE: The color is the color that the shading will be on the screen when the line of sight is displayed on the map.

    d. Select the box next to the Show Segment Profiles. The system will display the Segment Profile dialog box showing the line of sight for that particular segment. To show the next segment profile, just select the right arrow underneath the Show Segment Profile.
    e. Select the box to the left of the Show Segment Profile to uncheck it. The Segment Profile dialog box will disappear.
    f. Select the Execute button. The system will re display the route with red for the areas that have difficulty due to the side slope and Green in areas that are Passable.
    g. Select the Show Segment Details button. The system will display the Segment Details dialog box on top of the Analyze Route dialog box.

NOTES:

(1) If the dialog box is in the way of the Analyze Route dialog box and the route itself then left click on the dark gray area of the dialog box and drag it out of the way. The user should be able to have both the Segment Details and the Analyze Route dialog boxes on the screen and still view part of the route.

(2) The Segment Details dialog box will appear with the side slopes and grids of the difficulty areas of the first segment unless the first segment is passable without difficulty. If the segment is passable without difficulty then the Segment Details dialog box will display Segment is passable. If the Segment is not passable or has difficulty then the dialog box will display the side slope percentage and the grid of each area that has some difficulty.

(3) If the Segment Details dialog box has grids and percentages displayed in it, then the user can highlight the grid and a white dot will appear on the route to show the user where the difficult area is.

    h. Select the first grid on the Segment Detail dialog box. The system will display a white circle on the route where the difficult area is. Each grid that the user highlights, it will display the white circle on that grid.

NOTE: If the first segment is passable, then select the right arrow on the Analyze Route dialog box under the Show Segment Details until the Segment Details dialog box displays grids and percentages of the difficult areas.

    i. Select the Show All button on the Segment Details dialog box. The system will display the white circles on all the grids for that particular segment of the route that are difficult.
    j. Select the close button. The system will close the Segment Details dialog box and return to the Analyze Route dialog box as long as the user didn't close it earlier.
    k. Select the right arrow under the Show Segment Profile to move to the next waypoint.

NOTE: The number should have changed in between the arrows. It should show a number with a / and another number representing how many waypoints that are in the whole route. Each time the user select the right arrow, The number on the left of the / should change to the number of the next waypoint.

Performance Steps

    l. Select the Show Segment Details button again. The Segment Details dialog box will display with the side slopes and grids of the next segment.

NOTE: Again the Segment Details dialog box will display either Segment is Passable or it will display all the side slope percentages and grids of each point on the route that has some difficulty in travel.

    m. Select the close button. The system will return to the Analyze Route dialog box.

8. Select the Circular Line of Sight button on the bottom of the Analyze Route dialog box. The system will display the Circular Line of Sight dialog box.

NOTE: The Circular Line of Sight button allows the user to analyze a 360-degree line of sight from any point on the map. The user can control the analysis by entering the desired radius of the circle, the "Spacing" (distance between center points along a route), and the "Vertical Offset" (height from the ground to the desired eye level at the circle center) in the appropriate text box.

    a. Select the Points along a route button if the user wants to check the circular line of sight along the route, otherwise select the select points from map button to check the circular line of sight anywhere on the map.

    b. Select the Virtual Keypad next the Line of Sight Radius (m) text box and use the mouse pointer to enter the radius in meters between 1 and 10000 or place the cursor inside the text box and enter the number from the keyboard.

NOTE: Do the same for the Spacing (m) and the Vertical Offset (m) text boxes. If the user needs to know what the limits are that can be entered, the user can select the virtual keypad and look at the bottom of the keypad to see the limit.

    c. Select the Execute button. The system will display the Circular line of Sight on the first waypoint.

NOTE: If the user chose Points along route, the user would be able to just select the right arrow on the Circular line of Sight dialog box and the system would display the circular line of sight on the next way point. If the user chose Select points from map, the user can choose any point on the map or route.

    d. Select the Close button. The system will display the Analyze Route dialog box.

9. Select the Travel Time button. The system will display the Travel time dialog box.

NOTES:

(1) The Travel Time button allows the user to compute the time of each segment and the entire route based on the average speed factor.

(2) The Travel Time dialog box shows the Route length and the first segment length and the Estimated time under each that is blank until the user enters Average speed in Kilometers per hour and selects the Execute button.

    a. Select the Virtual keypad. Enter the speed between 1 and 200 by using the mouse pointer and selecting the OK button or enter the speed by placing the cursor in the Avg Speed (Kph) text box and type it in with the keyboard.

    b. Select the Execute button. The system will display the Estimated times under the Route Data and Segment Data.

    c. Select the Minimize button. The system will display the Minimize dialog box over the Analyze Route dialog box with the Travel Time dialog button with arrows facing left and right and two numbers with a / between them. The left number representing the waypoint that the data is displayed for and the number on the right representing the total number of waypoints.

    d. Select the Right arrow. The number on the left of the / should change to 2 and the number on the right should stay the same.

Performance Steps

  e. Select the Travel Time Dialog button. The system will display the Travel Time dialog box with the new Estimated times for the second segment.

NOTE: The user will need to select Minimize each time that the user wants to move to the next way point then select the Travel Time dialog button to see the Estimated times.

  f. Select the Close button. The system will return to the Analyze Route Dialog box.

10. Select the Edit Route button. The system will display the Map with the Edit Route Function bar on the right of the map and the route highlighted in blue with the last way point highlighted with a white circle.

NOTE: The Edit Route Button allows the user to edit the route on the map graphically. It also allows the user to grab, zoom, and view draw description, delete, and target a location route by name.

  a. Place the cursor at a spot on the map after the last way point and left click the mouse. The system will add a segment to the route. Each time the user selects a spot on the map, it will add a segment to the route.

NOTE: If the user added a way point on the map by mistake, just select the delete button and the system will delete the last segment added to the map and each time the delete button is selected, it will delete a segment of the route, Or select a way point on the route and move it to the desired location.

  b. Select the OK button. The system will return to the Analyze Route dialog box.
  c. Select the Close button. The system will display the Navigation dialog box.

11. Select the Manage button. The system will display the Edit Route dialog box.

NOTE: The Manage button allows the user to edit, rename, copy, or delete the route of march.

  a. Place the cursor in the Name text box and type a new name for the route using the keyboard and select the Rename button. The system will display the new name in the Name text box, or select the Virtual Keypad and enter the new name using the Mouse pointer and select the OK button. The name will display in the Name text box.

NOTE: To copy the route and add a new name, the user would type a new name in the Name text box and select the Copy button. The system would save a copy of the route with the new name in the Navigation dialog box. To delete a route, the user would type the name of the route that needs to be deleted and select the Delete button. The system would delete the route.

  b. Select the Map button. The system will display the Edit Route function bar with the route displayed on the map.

NOTE: From here the user can add or delete segments of the route.

  c. Select the OK button. The system will return to the Edit Route dialog box.
  d. Select the List button. The system will display the Edit way points dialog box with all the grids to the waypoints in the route.

NOTE: From here the user can add, modify, and delete waypoints and save the route.

  e. Scroll down and select the last way point in the list and then select the delete button. The system will delete the last waypoint.
  f. Select the Save button. The system will redisplay the route with the last way point deleted and return to the Edit Route dialog box.
  g. Select the Close button. The system will return to the Navigation dialog box.

12. Select the Center On button. The system will center on the route.
  a. Select the Msg Addressing button. The system will display the Message Addressing dialog box.

Performance Steps

NOTE: Refer to Task 171-147-0005 to address and send the route.

    b. Select the Close button on the Navigation dialog box. The system will return to the Ops screen.

Performance Measures	**GO**	**NO GO**
1. Select the NAV button.	——	——
2. Select the Route Tab if it is not already selected.	——	——
3. Select the Manage button.	——	——
4. Select the List button.	——	——
5. Select the Reverse button.	——	——
6. Select the Route Attributes button.	——	——
7. Select the Analyze Route button.	——	——
8. Select the Circular Line of Sight button on the bottom of the Analyze Route dialog box.	——	——
9. Select the Travel Time button.	——	——
10. Select the Edit Route button.	——	——
11. Select the Manage button.	——	——
12. Select the Center On button.	——	——

Evaluation Guidance: Score the Soldier GO if all steps are passed. Score the Soldier NO-GO if any step is failed. If the Soldier scores NO-GO, show him what was done wrong and how to do it correctly.

STP 11-25C13-SM-TG

EMPLOY QUICK SEND FUNCTIONS USING FBCB2 VERSION 3.4
171-147-0024

Conditions: Given a vehicle with an operational Force XXI Battle Command Brigade and Below (FBCB2) system with software version 3.4 loaded, precision lightweight global positioning system receiver (PLGR), enhanced position location radio system (EPLRS) if equipped, and single-channel ground air radio system (SINCGARS) with Internet controller (INC).

Standards: As a minimum you must: Employ screen operations functions to manage and exchange information, to include: Applying a message to the Quick Send button, naming the Button label, and naming the Balloon label.

Performance Steps

NOTE: The operator must create, and save a message to associate this button to before this task can be accomplished.

1. Select the Quick Send button. The Quick Send Button Setup dialog box will display.

NOTE: The Quick Send button allows the user to send a pertinent message in an expedient manner. The operator creates, and saves the message, then associates it with the Quick Send button. The Quick Send button has three dashes on it (---) and is located to the right of the NAV button.

2. Click the "+" sign to the left of the folder that contains the message that the user wants to associate to the Quick Send button. The will open and display all the messages in the folder.

3. Left-click on the message that will be associated to the Quick send button. The message will highlight.

4. Select the Button label text box and type a three-letter abbreviation that you want to apply to the button. Or select the virtual keyboard and type the abbreviation.

NOTE: The abbreviation is what you will see on the Quick Send button on the Function Bar.

5. Select the Balloon Label text box and type a short description of what the message is about, or select the virtual keyboard and type the description.

NOTE: The Balloon label is what the user will see when the cursor is placed over the Quick Send button; it will display what you typed in the Balloon Label text box.

6. Select the Apply button. The three-letter abbreviation will appear on the Quick Send button.

NOTE: Notice that the three letters that you typed in the Button Label text box appears on the Quick send Button on the function bar. If the user did not highlight a message to apply to the button, a selection error message would appear stating, "No message has been associated with this message."

7. Select the Display button. The system will display the message.

NOTE: From here the user can change the message address, forward the message, and save the message.

8. Select the Close button. The system will return to the Quick Send Button Setup dialog box.

9. Select OK. The system will return to the Ops screen.

NOTES:

(1) Place the cursor over the Quick Send Button for a couple of seconds and the short message that the user typed in the Balloon Label text box will appear.

STP 11-25C13-SM-TG

Performance Steps

(2) Send the message now by selecting the button.

Performance Measures	**GO**	**NO GO**
1. Select the Quick Send button.	——	——
2. Click the "+" sign to the left of the folder that contains the message that the user wants to associate to the Quick Send button.	——	——
3. Left-click on the message that will be associated to the Quick send button.	——	——
4. Select the Button label text box and type a three-letter abbreviation that you want to apply to the button.	——	——
5. Select the Balloon Label text box and type a short description of what the message is about, or select the virtual keyboard and type the description.	——	——
6. Select the Apply button.	——	——
7. Select the Display button.	——	——
8. Select the Close button.	——	——
9. Select OK.	——	——

Evaluation Guidance: Score the Soldier GO if all steps are passed. Score the Soldier NO-GO if any step is failed. If the Soldier scores NO-GO, show him what was done wrong and how to do it correctly.

STP 11-25C13-SM-TG

EMPLOY FILTERS FUNCTIONS USING FBCB2 VERSION 3.4
171-147-0025

Conditions: Given a vehicle with an operational Force XXI Battle Command Brigade and Below (FBCB2) system with software version 3.4 loaded, precision lightweight global positioning system receiver (PLGR), enhanced position location radio system (EPLRS) if equipped, and single-channel ground air radio system (SINCGARS) with Internet controller (INC).

Standards: As a minimum you must: Employ screen operations functions to manage and exchange information, to include: Select the desired radio button for the Labels, Friendly, Enemy, Unknown, and Georeference for the SA tab. Select Units to Collapse or Expand. Select the desired radio button for the labels for the Overlays tab. Load and Unload an overlay. Display and Hide an Overlay.

Performance Steps

NOTE: The FBCB2 system maintains near real-time data for friendly/enemy/unknown unit positions, and targets. The FBCB2 system also maintains data for fixed Geo-references such as bridges, mountains, and rivers. The "Filters" function permit the user to view or filter out friendly and enemy platforms or units according to currency, dimension, unit type, and echelon. Additionally, it loads, unloads, and displays overlays. This function also provides the user with the capability to monitor Geo-references and to filter overlays. Unit Standard Operating Procedures (SOP) determines filter use. Before this task can be completed, one or more of the following types of messages must be sent to the users system first: Spot Report, Obstacle Report, Bridge Report, Position Report, NBC1 Report. An Obstacle Overlay and one of any other type of Overlay should also be sent to the user's system. These reports generate the graphical images necessary to use all the Filter functions. Once the Messages and Overlays have been received, they must be saved to the user's system. Once the messages have been saved, the overlays can be displayed by opening the overlays and check the Keep Overlay displayed check box.

1. Select the Filters button on the Function bar or the F2 button on the keyboard. The system will display the Filters dialog box.

NOTE: The "Filters" function permit the user to view or filter out friendly and enemy platforms or units according to currency, dimension, unit type, and echelon. Additionally, it loads, unloads, and displays overlays. This function also provides the user with the capability to monitor Geo-references and to filter overlays. Unit Standard Operating Procedures (SOP) determines filter use. The Filters dialog box contains four tabs: SA (Situational Awareness), Collapse/Expand, Overlays, and Obstacle Overlays. If the user moves the Filters dialog box to the side, the user can view the labels, icons, and symbols appear and disappear from the screen as some of the functions are selected.

2. Select the SA tab if it is not already selected. The system will display the SA tab group.

NOTE: The SA tab group allows the user to access and filter SA data with the following functions: (All On), which displays all of the SA data that has been sent or received on the SA display area; (All Off), which filters out everything on the SA display area except the map, overlays, overlay objects and the platform's own position; (Labels), which filters the labels that are attached to the SA data; (Friendly), which enables the user to display or hide designated friendly platforms; (Enemy), which enables the user to display or hide designated enemy platforms; (Unknown), which displays or hides unidentified units; and Georef, which displays or hides geographical references.

    a. Select the None radio button under the Labels field. The labels for all the SA on the map will disappear.

Performance Steps

NOTE: Notice how the word SET highlighted in yellow appeared on the Filters button on the function bar. This is to inform the user that filters have been set. By selecting the All button, the labels will reappear.

    b. Select the Select radio button under the Friendly field. The Currency, Dimension, Type, and Echelon fields will display.

NOTE: These functions allow the user to display or hide friendly SA according to their Current, Stale or Old settings, and Air or Ground units, or by Branch or affiliation, and by type of Echelon. The user can uncheck any field that he or she does not want to see.

    c. Select the All button under the Friendly field. The all friendly SA will reappear on the map.
    d. Select the Select radio button under the Enemy field. The Currency, Dimension, Type, and Source of Info fields will appear.

NOTE: The only difference between Friendly and Enemy is Echelon was replaced with Source of Info, which allows the user to filter enemy icons based on its source. The Three sources are: ASAS (All Source Analysis System), FAAD (Forward Area Air Defense), and Spot Rprt (Spot Report). Deselect the check box of the source that the user does not want. Users should avoid having both the ASAS and Spot Rprt checked at the same time. This can cause a distorted picture because enemy icons may be depicted twice.

    e. Select the All radio button under the Enemy field. The system will redisplay all the enemy icons.
    f. Select the None radio button under the Unknown field. Al of the Unidentified icon will disappear off the map.
    g. Select the All radio button under the Unknown field. The system will display all the unidentified icons.
    h. Select the None radio button under the Georef field. All of the Geographical references will disappear.
    i. Select the All radio button under the Georef field. The system will redisplay the Geographical Geo-references.

  3. Select the Collapse/Expand tab. The system will display the Collapse/Expand tab group.

NOTE: This function allows the user to collapse and expand the elements of a combat unit on the local display to a single Center of Mass (CM). This function collapses multiple unit icons under a single unit icon. Units collapse in accordance with the current UTO.

    a. Select the Search text box by placing the cursor in the text box and left clicking. The text box will become active.

NOTE: The Search text box is used as an expedient way of finding a unit to expand or collapse.

    b. Type the name of the unit that needs expanding or collapsing.

NOTE: The name must be spelled and spaced exactly the way it is in the UTO in order for the search function to find the unit. If the name was spelled and spaced right, the name will be highlighted when the search button is pushed.

    c. Select the "+" or "-" sign next to the unit.

NOTE: Selecting the "-" sign expands the unit selected into specific level. Selecting the "+"sign collapses the unit into one icon.

  4. Select the Overlays tab. The system will display the Overlays tab group.

NOTE: This function allows the user to display or hide labels on the Overlay and Load and Unload Overlays.

    a. Select the None radio button under the Labels field. All the Labels that are on the overlays will disappear.

STP 11-25C13-SM-TG

Performance Steps

    b. Select the All radio button under the Labels field. The system will redisplay the Labels.

    c. Select the Selected radio button under the Overlays field. The system will display the names of all the overlays that have been created, received, and saved on the system along with the Load, Unload Selected, and Unload All buttons.

NOTE: The Load button will load the overlays that have been check marked. The Unload Selected will unload all the overlays that have been check marked. The Unload All button will unload all overlays that have been saved on the system. If the user received any overlays from other platforms, they must be saved to a folder on the user's computer before the system will load them. The overlays that the user creates and saves can also be loaded and unloaded.

    d. Select the Load button. The system will display the Overlay Loader dialog box with all the folders that have been created on the system.

    e. Select the folder that has the overlay(s) that the user wants to load by double-clicking the folder or selecting the "+" sign. The system will highlight the overlay name.

    f. Select the OK button. The system will return to the Filters dialog box with the name of the overlay displayed.

NOTE: If there are more overlays that the user wants to load, then the user will have to do the same steps to load them.

    g. Select the overlay that was just loaded by left clicking on the name. The overlay name will be highlighted.

    h. Select the Unload Selected button. The system will display the Unload Selected Overlay dialog box stating "You are about to unload overlay with the 'name' Continue with this action? with an OK button.

    i. Select the OK button. The system will unload the overlay and return to the Overlays tab.

5. Select the Obstacle Overlays tab. The system will display the Obstacle Overlays tab group with an All and None radio button and the names of the Originators and DTG of all the Obstacle Overlays that have been created, received, and saved on the system.

NOTE: An Obstacle overlay must be loaded in order to view this function.

    a. Select the None button. All the Obstacle Overlays that are displayed on the map will disappear.

    b. Select the All radio button. The Obstacle overlays will reappear on the map.

    c. Select an Obstacle Overlay by highlighting the Originator. The name of the Originator will highlight.

    d. Select the Delete Selected button. The system will remove the Obstacle Overlay from the Obstacle Overlays tab group but not from the system.

    e. Select the Close button. The system will return to the Ops screen.

Performance Measures	GO	NO GO
1. Select the Filters button on the Function bar or the F2 button on the keyboard.	——	——
2. Select the SA tab if it is not already selected.	——	——
3. Select the Collapse/Expand tab.	——	——
4. Select the Overlays tab.	——	——
5. Select the Obstacle Overlays tab.	——	——

Evaluation Guidance: Score the Soldier GO if all steps are passed. Score the Soldier NO-GO if any step is failed. If the Soldier scores NO-GO, show him what was done wrong and how to do it correctly.

Skill Level 2

Subject Area 9: COMBAT COMMUNICATIONS PLANNING

Direct Implementation of an FM Voice Data Communications Network
113-587-7133

Conditions: Given the unit's OPORD, unit's SOI, unit's tactical SOP, frequency modulated (FM) radio installation kits, FM radios, COMSEC fill devices, COMSEC devices, extended range antennas, DA PAM 738-750, applicable technical manuals, FM 24-18, and FM 24-33.

Standards: The FM NCS received a radio check from all stations.

Performance Steps

1. Review the unit's OPORD
   a. Identify the mission of the FM net.
   b. Identify the available assets.
   c. Identify the transmission plan diagram.
   d. Identify the locations of the stations within the net.
   e. Identify the need for radio RETRANS.

2. Allocate the resources.
   a. Ensure users have the required radio equipment to operate in the net.
   b. Ensure users have the unit's current SOI (manual or electronic).
   c. Ensure the SOI database (electronic) is modified and verified down to the lowest user.
   d. Ensure users have the required current COMSEC key.
   e. Ensure users have the required COMSEC fill device for their radio/COMSEC equipment.
   f. Ensure RETRANS operators have all equipment/supplies outlined in the unit's tactical SOP.

3. Establish the radio net.
   a. Ensure users conduct a proper PMCS of their assigned equipment.
   b. Advise the unit commander on the status of the equipment.
   c. Ensure users are notified of the net start-up time.
   d. Ensure the net is open.
   e. Ensure the proper radio procedures are being utilized.
   f. Ensure radio checks are conducted with all the stations.

Performance Measures   <u>GO</u>   <u>NO GO</u>

1. Reviewed the unit's OPORD.   ___   ___
   a. Identified the mission of the FM net.
   b. Identified the available assets.
   c. Identified the transmission plan diagram.
   d. Identified the locations of the stations within the net.
   e. Identified the need for radio RETRANS.

2. Allocated the resources.   ___   ___
   a. Ensured users had the required radio equipment to operate in the net.
   b. Ensured users had the unit's current SOI (manual or electronic).
   c. Ensured the SOI database was modified and verified down to the lowest user.
   d. Ensured users had the required COMSEC fill device for their radio/COMSEC equipment.
   e. Ensured users had an extended range antenna if required at their location.
   f. Ensured the RETRANS operators had all the equipment/supplies outlined in the unit's tactical SOP.

STP 11-25C13-SM-TG

Performance Measures   <u>GO</u>   <u>NO GO</u>

   3. Established the radio net.   ——   ——
      a. Ensured users conducted a proper PMCS of their assigned equipment.
      b. Advised the unit commander on the status of the equipment.
      c. Ensured users were notified of the net start-up time.
      d. Ensured the net was opened.
      e. Ensured the proper radio procedures were utilized.
      f. Ensured that radio checks were conducted with all the stations.

Evaluation Guidance: Score the Soldier a GO if all PMs are passed. Score the Soldier a NO-GO if any PM is failed. If the Soldier fails any PM, show what was done wrong and how to do it correctly. Have the Soldier perform the PMs until they are done correctly.

References
    Required   **Related**
    DA PAM 738-750   FM 11-32
    EQUIPMENT TM(S)
    FM 24-18
    FM 24-33
    UNIT OPORD
    UNIT SOI
    UNIT TACTICAL SOP

STP 11-25C13-SM-TG

Direct Implementation of a Single-Channel Tactical Satellite Communications Network
113-589-7120

Conditions: Given the unit's OPORD, unit's SOI, unit's tactical SOP, TACSAT communications equipment, COMSEC fill devices, COMSEC devices, applicable technical manuals, DA PAM 738-750, FM 24-11, FM 24-18, and FM 24-33.

Standards: The TACSAT communications NCS received a radio check from all the stations.

Performance Steps

1. Review the unit's OPORD.
    a. Identify the mission of the TACSAT net.
    b. Identify the transmission plan diagram.
    c. Identify the azimuth and elevation of the satellite to be used.
    d. Identify the available assets, to include compatibility of different TACSAT systems.
    e. Identify the NCS.
    f. Identify the locations of the stations within the net.

2. Allocate the resources.
    a. Ensure users have the required TACSAT equipment to operate in the net.
    b. Ensure users have the unit's current SOI (manual or electronic) or frequency information to include uplink, downlink, and offset.
    c. Ensure users have the required current COMSEC key.
    d. Ensure users have the required COMSEC fill device for their radio/COMSEC device.
    e. Ensure users have loaded the presets for COMSEC devices.
    f. Ensure users have the correct configuration and have loaded the terminal.
    g. Ensure users have the correct azimuth and elevation to be used from their location.

3. Establish the TACSAT net.
    a. Ensure users conduct a PMCS of their assigned equipment.
    b. Advise the commander on the status of the equipment.
    c. Ensure users are notified of the net start-up time.
    d. Ensure the net is open.
    e. Ensure the proper radio procedures are being utilized.
    f. Ensure radio checks are conducted with all the stations.

Performance Measures   <u>GO</u>   <u>NO GO</u>

1. Reviewed the unit's OPORD.   ——   ——
    a. Identified the mission of the TACSAT net.
    b. Identified the transmission plan diagram.
    c. Identified the azimuth and elevation of the satellite to be used.
    d. Identified the available assets, to include compatibility of different TACSAT systems.
    e. Identified the NCS.
    f. Identified the locations of the stations within the net.

2. Allocated the resources.   ——   ——
    a. Ensured users had the required TACSAT equipment to operate in the net.
    b. Ensured users had the unit's current SOI (manual or electronic) or frequency information to include uplink, downlink, and offset.
    c. Ensured users had the required current COMSEC key.
    d. Ensured users had the required COMSEC fill device for their radio/COMSEC equipment.
    e. Ensured all presets were correct in the COMSEC equipment.
    f. Ensured users had the correct configuration and had loaded the terminal.

16 February 2005

STP 11-25C13-SM-TG

Performance Measures	**GO**	**NO GO**
g. Ensured users had the correct azimuth and elevation to be used from their location.		
3. Established the radio net.	___	___
a. Ensured users conducted a proper PMCS of their assigned equipment.		
b. Advised the unit commander on the status of the equipment.		
c. Ensured users were notified of the net start-up time.		
d. Ensured the net was opened.		
e. Ensured the proper radio procedures were utilized.		
f. Ensured that radio checks were conducted with all the stations.		

Evaluation Guidance: Score the Soldier a GO if all PMs are passed. Score the Soldier a NO-GO if any PM is failed. If the Soldier fails any PM, show what was done wrong and how to do it correctly. Have the Soldier perform the PMs until they are done correctly.

References

    Required   **Related**
    DA PAM 738-750
    EQUIPMENT TM(S)
    FM 24-11
    FM 24-18
    FM 24-33
    UNIT OPORD
    UNIT SOI
    UNIT SOP

Select Team Radio Site
113-611-1001

Conditions: Given an operation order (OPORD), unit standard operation procedures (SOP), map of the area, protractor, lensatic compass, a minimum of two operational radios, FM 24-18 and FM 11-487-1 (if available) in a tactical or nontactical situation select a team radio site.

NOTE: Supervision and assistance are available.

Standards: Selected a team radio site and the Soldier received a GO in all areas.

Performance Steps

1. Determine the requirements in accordance with the OPORD.
    a. Technical requirements.
        (1) Maximum signal-to-noise ratio at the receiver site.
        (2) Maximum effective power radiated in the desired direction from the transmitter site.
    b. Tactical requirements.
        (1) Local command requirements.
        (2) Cover and Concealment.
            (a) Determine if there are any terrain features that can be used to your unit's advantage.
            (b) Determine if any material cover and/or concealment exists.
        (3) Practical considerations.
            (a) Antenna siting.
            (b) Remote operations (if applicable).
            (c) Antenna concealment.
            (d) Camouflage.
        (4) Local communications.

NOTE: Ensure contact can be maintained between the radio site and the serviced headquarters or communications center at all times.

            (a) Field telephone.
            (b) Messenger.
    c. General site requirements.
        (1) Availability.
        (2) Suitability.
        (3) Accessibility.
        (4) Security.

2. Recon the map to determine requirements.
    a. Plot all known grid coordinates.
    b. Determine the distance between sites.
    c. Select a site for primary and alternate site.

Performance Measures	GO	NO GO

1. Determined the requirements in accordance with the OPORD.
    a. Technical requirements.
    b. Tactical requirements.
    c. General site requirements.
	——	——

2. Reconed the map to determine requirements.
    a. Plotted all known grid coordinates.
    b. Determined the distance between sites.
    c. Selected a site for primary and alternate site.
	——	——

Evaluation Guidance: Score the Soldier a GO if all PMs are passed. Score the Soldier a NO-GO if any PM is failed. If the Soldier fails any PM, show what was done wrong and how to do it correctly. Have the Soldier perform the PMs until they are done correctly.

References

Required	**Related**
FM 11-487-1	
FM 24-18	
UNIT OPORD	
UNIT SOP	

STP 11-25C13-SM-TG

Prepare Input to Signal Annex (Paragraph 5) of the OPORD
113-611-6112

Conditions: Given an incomplete OPORD, FM 24-16, FM 105-5, unit's SOI, mission statement, and a map of the operational area.

Standards: Prepared the input to the Signal Annex (paragraph 5), identified the battalion's communications-electronics responsibility to the mission, and the supervisor approved the input information.

Performance Steps

1. Review supporting documents.
    a. Determine the mission.
    b. Determine the communications-electronics responsibility to the mission. (Review the unit's OPORD.)

2. Prepare input for the applicable portion of the unit's SOI to be in effect. Prepare any special instructions relating to signal matters.

3. List the locations of the command posts (CPs).
    a. List CP location for the issuing unit.
    b. List CP locations for subordinate units.
    c. List CP location for the command group.

4. Determine if information on future locations of major headquarters is needed. List all future signal system planning and location requirements as necessary.

5. Analyze the mission as it pertains to the communications-electronics responsibility.
    a. Ensure all input information to the communications-electronics of the Signal Annex is listed.
    b. Identify the communications-electronics responsibility to the mission.

6. Submit input information for supervisor's approval.

Performance Measures	**GO**	**NO GO**
(Refer to FM 101-5, FM 24-16, unit's SOI, and mission statement for PMs 1 through 5.)		

1. Reviewed supporting documents.
   a. Determined the mission.
   b. Determined the communications-electronics responsibility to the mission. (Reviewed the unit's OPORD.)
	——	——

2. Prepared input for the applicable portion of the unit's SOI to be in effect. Prepared any special instructions relating to signal matters.	——	——

NOTE: Refer to the map for PMs 3 through 5.

3. Listed the locations of the CPs.
   a. Listed CP location for the issuing unit.
   b. Listed CP locations for subordinate units.
   c. Listed CP location for the command group.
	——	——

4. Determined if information on future locations of major headquarters was needed. Listed all future signal system planning and location requirements as necessary.	——	——
5. Analyzed the mission as it pertained to the communications-electronics responsibility. a. Ensured all input information to the communications-electronics of the Signal Annex was listed.	——	——

STP 11-25C13-SM-TG

Performance Measures   <u>GO</u>   <u>NO GO</u>

    b. Identified the communications-electronics responsibility to the mission.

6. Submitted input information for supervisor's approval.   ——   ——

Evaluation Guidance: Score the Soldier a GO if all PMs are passed. Score the Soldier a NO-GO if any PM is failed. If the Soldier fails any PM, show what was done wrong and how to do it correctly. Have the Soldier perform the PMs until they are done correctly.

References
Required	**Related**
FM 101-5	
FM 24-16	
UNIT SOI	

STP 11-25C13-SM-TG

Direct Implementation of a High Frequency (HF) Communications Network
113-620-7089

Conditions: Given the unit's OPORD, unit's SOI, unit's tactical SOP, HF equipment, COMSEC fill devices, DA PAM 738-750, applicable technical manuals, FM 11-32, FM 24-18, and FM 24-33.

Standards: Installed the HF communications net IAW the unit's OPORD; and the NCS received radio checks from all the stations.

Performance Steps

1. Review the unit's OPORD.
    a. Identify the mission of the HF net.
    b. Identify the transmission plan diagram.
    c. Identify the available assets.
    d. Identify the NCS.
    e. Identify the locations of the stations within the net.
    f. Identify the need for a RETRANS.

2. Allocate the resources.
    a. Ensure users have the required equipment to include all necessary antenna sets and have an understanding of which antenna to use in each situation.
    b. Ensure users have the unit's current SOI (manual or electronic).
    c. Ensure users have the current required COMSEC key.
    d. Ensure users have the required COMSEC fill device for their type of radio set.

3. Establish the radio net.
    a. Ensure users conduct a proper PMCS of their assigned equipment.
    b. Advise unit commander on the status of the equipment.
    c. Ensure users are aware of the net start-up time.
    d. Ensure the net is open.
    e. Ensure the proper radio procedures are being utilized.
    f. Ensure radio checks are conducted with all the stations.

Performance Measures	**GO**	**NO GO**

1. Reviewed the unit's OPORD.
   a. Identified the mission of the net.
   b. Identified the transmission plan diagram.
   c. Identified the available assets.
   d. Identified the NCS.
   e. Identified the locations of the stations within the net.
   f. Identified the need for a RETRANS.
	——	——

2. Allocated the resources.
   a. Ensured users had the required equipment.
      (1) Ensured users had all the needed antenna equipment.
      (2) Ensured users knew which antenna to use in each situation.
   b. Ensured users had the unit's current SOI (manual or electronic).
   c. Ensured users had the current required COMSEC key.
   d. Ensured users had the required COMSEC fill device for their type of radio set.
	——	——

3. Established the radio net.
   a. Ensured users conducted the proper PMCS of their assigned equipment prior to establishing net.
   b. Advised unit commander of the status of all required equipment.
	——	——

Performance Measures <u>GO</u>　　<u>NO GO</u>
    c. Ensured users were aware of net start-up time.
    d. Ensured the net was opened on time.
    e. Ensured the proper radio procedures were utilized.
    f. Ensured that radio checks were conducted with all the stations.

Evaluation Guidance: Score the Soldier a GO if all PMs are passed. Score the Soldier a NO-GO if any PM is failed. If the Soldier fails any PM, show what was done wrong and how to do it correctly. Have the Soldier perform the PMs until they are done correctly.

References
    Required　　　　　　　　　　　　　　**Related**
    DA PAM 738-750
    EQUIPMENT TM(S)
    FM 11-32
    FM 24-18
    FM 24-33
    UNIT OPORD
    UNIT SOI
    UNIT TACTICAL SOP

STP 11-25C13-SM-TG

Subject Area 12: CONSTRUCT ANTENNAS

Construct Field Expedient Antennas
113-596-1098

Conditions: Given WD-1, materials for an insulator, spacing sticks, rope, wire or tape, and FM 24-18 in a field environment construct a field expedient antenna.

Standards: The Soldier received a GO on all five steps and constructed an antenna and established communications.

Performance Steps

1. Determine length of elements for the frequency being used.
    a. Use a quick-reference chart to determine the length of the elements for the frequency you will be using.
    b. Cut these elements from WD-1/TT field wire (or similar wire).

NOTE: The best wire for antennas is copper or aluminum. In an emergency use any wire available.

    c. Cut spacing sticks to equal length.

2. Assemble the antenna.
    a. Place the end of the sticks together to form a triangle and tie the ends with wire, tape, or rope.
    b. Attach an insulator to each corner.
    c. Attach a ground-plane wire to each insulator.
    d. Bring the other ends of the ground-plane wires together, attach them to an insulator and tie securely.
    e. Strip about 3 inches of insulator from each wire and twist them together.
    f. Tie one end of the radiating element wire to the other side of insulator and the other end to another insulator.
    g. Strip about 3 inches of insulation from the radiation element at insulator.
    h. Cut enough WD-1/TT field wire to reach from the proposed location to the antenna to the radio set.

NOTE: Keep this line as short as possible: excess length reduces efficiency of system.

    i. Tie a knot at each end of this cable pair to keep it from unraveling.
    j. Identify one wire in the pair and tie a knot at each end to identify it as "hot" lead.
    k. Remove insulation from the "hot" wire and tie it to the radiation element wire at insulator.
    l. Remove insulation from the other wire and attach it to the bare ground-plane element wires at insulator.
    m. Tape all connections and do not allow the radiating element wire to touch the ground-plane wires.

3. Erect the antenna.
    a. Attach a rope to the insulator on the free end of the radiating element and toss the rope over the branches of a tree.
    b. Pull the antenna as high as you can, keeping the lead-in routed down through the triangle.
    c. Secure the rope to hold the antenna in place.

4. Attach antenna to radio.
    a. At the radio set, remove about 1 inch of insulation from the "hot" lead and about 3 inches of insulation from the other wire.
    b. Attach the "hot" line to the antenna terminal (doublet connector, is so labeled.)
    c. Attach the other wire to the metal case-the handle, for example.

STP 11-25C13-SM-TG

Performance Steps

NOTE: Be sure both connections are tight or secure.

   5. Establish communications.

Evaluation Preparation: Setup: Ensure that all the information, references, and equipment required to perform the task are available. Use the TMs and the evaluation guide to score the Soldier's performance.

Brief Soldier: Tell the Soldier what he is required to do IAW the task condition and standards.

Performance Measures	**GO**	**NO GO**
1. Determined length of elements for the frequency being used.	——	——
2. Assembled the antenna.	——	——
3. Erected the antenna.	——	——
4. Attached antenna to radio.	——	——
5. Established communications.	——	——

Evaluation Guidance: Score the Soldier a GO if all PMs are passed. Score the Soldier a NO-GO if any PM is failed. If the Soldier fails any PM, show what was done wrong and how to do it correctly. Have the Soldier perform the PMs until they are done correctly.

References

   Required   **Related**
   FM 24-18   FM 24-19

Subject Area 13: AUTOMATION OPERATIONS

Troubleshoot Local Area Network (LAN)
113-580-0056

Conditions: Given a properly configured LAN unable to process data, cable-splicing kit, LAN circuit diagram, BFACS Software User's Manual (SUM), Computer's User's Manual (CUM), TM 11-5895-1348-13&P, TM 11-5895-1546-12&P, FM 24-7, and test a message.

Standards: The Soldier received a GO in all four steps and sent a test message and received an acknowledgement from all addresses within the network.

Performance Steps

1. Verify reported malfunction.
    a. Review operator's actions.
    b. Perform visual inspections.

2. Isolate malfunction.
    a. Determine inoperable terminal.
    b. Perform on-line diagnostics.
    c. Perform off-line diagnostics.

3. Repair or replace the LAN cable/connectors/terminators.

4. Perform an operational check.
    a. Send a test message to all addressees.
    b. Receive an acknowledgement from all addressees.

NOTE: If the operational check fails, refer to the troubleshooting section of appropriate computer manual(s).

Performance Measures	GO	NO GO
1. Verified reported malfunction.	——	——
2. Isolated malfunction.	——	——
3. Repaired or replaced the LAN cable/connectors/terminators.	——	——
4. Performed an operational check.	——	——

Evaluation Guidance: Score the Soldier a GO if all PMs are passed. Score the Soldier a NO-GO if any PM is failed. If the Soldier fails any PM, show what was done wrong and how to do it correctly. Have the Soldier perform the PMs until they are done correctly.

References

Required	Related
BFACS SUM	
CUM	
FM 24-7	
TM 11-5895-1348-13&P	
TM 11-5895-1546-12&P	

Skill Level 3

Subject Area 9: COMBAT COMMUNICATIONS PLANNING

Plan FM Voice and Data Communications Net
113-611-6002

Conditions: Given FM 3-25.26, FM 24-16, TM 11-5820-890-20-1, TM 11-5820-890-20-2, TM 11-5825-283-23, list of equipment (SINCGARS radio set, AN/CYZ-10, EPLRS radio set AN/VSQ-2(V)2), and unit's SOI, SOP, and operation plan (OPLAN).

Standards: Incorporated the voice and data communications net plan into the OPORD, and the commander approved the plan.

Performance Steps

1. Identify mission requirements of the FM voice and data communications net.
    a. Review the OPLAN.
    b. Review the assets.

2. Formulate signal estimate.
    a. Develop the transmission plan diagram.
    b. Identify the sustainment requirements.

3. Planning considerations.
    a. Terrain.
    b. Weather.
    c. Type of unit.
    d. Number of units.
    e. Availability of COMSEC devices.
    f. Restricted frequencies.
    g. Antenna requirements.
    h. Power requirements.
    i. Electronic warfare (EW) threats.

4. Submit the plan for approval.
    a. Brigade or battalion signal officer.
    b. Training operations S3.

5. Incorporate the approved plan into the OPORD.

Performance Measures	**GO**	**NO GO**

1. Identified mission requirements of the FM voice and data communications net.
    a. Reviewed the OPLAN.
    b. Reviewed the assets.
	——	——

2. Formulated signal estimate.
    a. Developed the transmission plan diagram.
    b. Identified the sustainment requirements.
	——	——

3. Planned considerations.
    a. Terrain.
    b. Weather.
    c. Type of units.
    d. Number of units.
    e. Availability of COMSEC devices.
    f. Restricted frequencies.
	——	——

Performance Measures	GO	NO GO
g. Antenna requirements.		
h. Power requirements.		
i. EW threats.		
4. Submitted the plan for approval.	——	——
a. Brigade or battalion signal officer.		
b. Training operations S3.		
5. Incorporated the approved plans into the OPORD.	——	——

Evaluation Guidance: Score the Soldier a GO if all PMs are passed. Score the Soldier a NO-GO if any PM is failed. If the Soldier fails any PM, show what was done wrong and how to do it correctly. Have the Soldier perform the PMs until they are done correctly.

References

Required	**Related**
FM 24-16	FM 24-22
FM 3-25.26	
TM 11-5820-890-20-1	
TM 11-5820-890-20-2	
TM 11-5825-283-23	
UNIT OPLAN	
UNIT SOI	
UNIT SOP	

STP 11-25C13-SM-TG

Plan a Single-Channel Tactical Satellite Communications Network
113-611-6004

Conditions: Given the unit's OPLAN, SOI, and SOP; list of TACSAT communications equipment; DA PAM 738-750; FM 11-24, FM 24-16, and FM 24-22; TM 11-5896-1181-20 and TM 11-5895-1180-20; and the AN/CYZ-10.

Standards: The commander approved the OPORD.

Performance Steps

1. Review the unit's SOP and SOI to determine specific TACSAT communications requirements.
    a. Determine the net requirements.
    b. Determine the locations of the communication sites as required.

2. Review the equipment list to determine the assets available to support the mission.

3. Determine the logistical support required to install and maintain the TACSAT communications network.
    a. Type of unit.
    b. Number of units.
    c. Availability of COMSEC devices.
    d. Restricted frequencies.
    e. Antenna requirements.
    f. Power requirements.

4. Review equipment capabilities.

5. Submit the plan for approval.
    a. Brigade or battalion signal officer.
    b. S3 operations.

6. Incorporate the approved plan into the OPORD.

Performance Measures	GO	NO GO

1. Reviewed the unit's SOP and SOI to determine specific TACSAT communications requirements.
    a. Determined the net requirements.
    b. Determined the locations of the communication sites as required.
	——	——

2. Reviewed the equipment list to determine the assets available to support the mission.	——	——

3. Determined the logistical support required to install and maintain the TACSAT communications network.
    a. Type of unit.
    b. Number of units.
    c. Availability of COMSEC devices.
    d. Restricted frequencies.
    e. Antenna requirements.
    f. Power requirements.
	——	——

4. Reviewed equipment capabilities.	——	——

5. Submitted the plan for approval.
    a. Brigade or battalion signal officer.
    b. S3 operations.
	——	——

STP 11-25C13-SM-TG

Performance Measures   <u>**GO**</u>   <u>**NO GO**</u>

   6. Incorporated the approved plan into the OPORD.   ——   ——

Evaluation Guidance: Score the Soldier a GO if all PMs are passed. Score the Soldier a NO-GO if any PM is failed. If the Soldier fails any PM, show what was done wrong and how to do it correctly. Have the Soldier perform the PMs until they are done correctly.

References
    Required   **Related**
    DA PAM 738-750   FM 3-25.26
    FM 11-24
    FM 24-16
    FM 24-22
    TM 11-5895-1180-20
    TM 11-5895-1181-20
    UNIT OPLAN
    UNIT SOI
    UNIT SOP

STP 11-25C13-SM-TG

Plan an HF Communications Network
113-611-6005

Conditions: Given the unit's OPORD/OPLAN, SOP, and SOI; the AN/CYZ-10; FM 11-65, FM 24-16, and FM 24-18.

Standards: Prepared and completed the HF communications network plan, and the commander approved the plan.

Performance Steps

1. Review the unit's current SOP and SOI to determine specific HF requirements.
    a. Determine the net requirements.
    b. Determine the locations of the communications sites as required.

2. Review the equipment and troop list to determine the assets available to support the mission.

3. Determine the logistical support required to install and maintain the HF communications system.

4. Review equipment capabilities.

5. Review the time permitted for installation, type of terrain, and expected weather conditions under which the system will operate.

6. Determine propagation mode(s).
    a. Ground wave.
    b. Skywave.

7. Request engineering support as required.
    a. Propagation prediction.
    b. Antenna sighting and orientation.
    c. Frequency.

NOTE: The planner may be required to do part or all of the systems engineering, utilizing the appropriate ground wave propagation chart and/or the intermediate and short-distance skywave propagation charts. Coordinate with the frequency management section on the frequency selection.

8. Determine the network layout as required.

9. Assign the NCS as required.

10. Assign frequencies and time periods for operations.

11. Prepare the radio net diagram.

12. Submit the plan for approval to battalion S3.

Performance Measures	**GO**	**NO GO**

1. Reviewed the unit's current SOP and SOI to determine specific HF requirements.
    a. Determined the net requirements.
    b. Determined the locations of the communications sites as required.
	——	——

2. Reviewed the equipment and troop list to determine the assets available to support the mission.	——	——
3. Determined the logistical support required to install and maintain the HF communications system.	——	——
4. Reviewed equipment capabilities.	——	——

Performance Measures	GO	NO GO
5. Reviewed the time permitted for installation, type of terrain, and expected weather conditions under which the system operated.	——	——

6. Determined propagation mode(s).
    a. Ground wave.
    b. Skywave.
	——	——

7. Requested engineering support as required.
    a. Propagation prediction.
    b. Antenna siting and orientation.
    c. Frequency.
	——	——

8. Determined the network layout as required.	——	——
9. Assigned the NCS as required.	——	——
10. Assigned frequencies and time periods for operations.	——	——
11. Prepared the radio net diagram.	——	——
12. Submitted the plan for approval to battalion S3.	——	——

Evaluation Guidance: Score the Soldier a GO if all PMs are passed. Score the Soldier a NO-GO if any PM is failed. If the Soldier fails any PM, show what was done wrong and how to do it correctly. Have the Soldier perform the PMs until they are done correctly.

References

Required	**Related**
FM 11-65	
FM 24-16	
FM 24-18	
UNIT OPLAN	
UNIT OPORD	
UNIT SOI	
UNIT SOP	

Subject Area 10: RADIO NET OPERATION

Inspect Station/Net Operation and Duties
113-571-7002

Conditions: Given ACP 124(D), ACP 125(E), ACP 126(C), AR 5-12, AR 380-5, DA PAM 738-750, FM 24-18, FM 24-33, TB 380-41, unit SOI, and inspection checklist.

Standards: Completed an inspection of the station/net operations (refer to checklist) and recorded and submitted results to proper authorities.

Performance Steps

1. Obtain and review appropriate references.
2. Check processing of incoming messages. (Refer to unit SOI and FM 24-18.)
3. Check processing of outgoing messages. (Refer to unit SOI and FM 24-18.)
4. Check station logs and forms. (Refer to FM 24-18, DA PAM 738-750, and TB 380-41.)
5. Check observance of signal security and physical security. (Refer to ACP 124(D), ACP 125(E), ACP 126(C), FM 24-18, and AR 380-5.)
6. Submit required reports. (Refer to FM 24-18, FM 24-33, AR 5-12, and unit SOP.)

Performance Measures	GO	NO GO
1. Obtained and review appropriate references.	——	——
2. Checked processing of incoming messages. (Refer to unit SOI and FM 24-18.)	——	——
3. Checked processing of outgoing messages. (Refer to unit SOI and FM 24-18.)	——	——
4. Checked station logs and forms. (Refer to FM 24-18, DA PAM 738-750, and TB 380-41.)	——	——
5. Checked observance of signal security and physical security. (Refer to ACP 124(D), ACP 125(E), ACP 126(C), FM 24-18, and AR 380-5.)	——	——
6. Submitted required reports. (Refer to FM 24-18, FM 24-33, AR 5-12, and unit SOP.)	——	——

Evaluation Guidance: Score the Soldier a GO if all PMs are passed. Score the Soldier a NO-GO if any PM is failed. If the Soldier fails any PM, show what was done wrong and how to do it correctly. Have the Soldier perform the PMs until they are done correctly.

References
    Required   **Related**
    ACP 124(D)
    ACP 125(E)
    ACP 126(C)
    AR 380-5
    AR 5-12
    DA PAM 738-750
    FM 24-18
    FM 24-33
    TB 380-41
    UNIT SOI

STP 11-25C13-SM-TG

Conduct Communications Security Inspections
113-573-1004

Conditions: Given ACP 124(D), ACP 125(E), ACP 126(C), AR 380-5, FM 24-18, FM 24-33, TB 380-41, unit SOI, and inspection checklist.

Standards: Completed an operations security inspection (appropriate per unit/mission) and recorded and submitted results to proper authorities.

Performance Steps

1. Check physical security. (Refer to TB 380-41.)
   a. Check defense of restricted areas (fences, guards, dogs, and so forth).
   b. Check access roster.
   c. Check doors, locks, and windows.
   d. Check SF 702 (Security Container Check Sheet), SF 700 (Security Container Information), DA Form 2653-R (COMSEC Account – Daily Shift Inventory [LRA]), and SF 153 (COMSEC Material Report).
   e. Check emergency evacuation.
   f. Check destruction records.
   g. Check the reporting of violation suspected or actual.

2. Check transmission security. (Refer to ACP 124(D), ACP 125(E), ACP 126(C), FM 24-18, and unit SOI.)
   a. Ensure 20-second transmission time is being enforced.
   b. Ensure messages are preplanned before transmitting.
   c. Ensure equipment is operated in the secure mode.

3. Check cryptographic security utilizing the crypto facility checklist in TB 380-41.

4. Submit completed results of the inspection to proper authorities.

Performance Measures	GO	NO GO

1. Checked physical security. (Refer to TB 380-41.)
    a. Checked defense of restricted areas (fences, guards, dogs, and so forth).
    b. Checked access roster.
    c. Checked doors, locks, and windows.
    d. Checked SF 702, SF 700, DA Form 2653-R, and SF 153.
    e. Checked emergency evacuation.
    f. Checked destruction records.
    g. Checked the reporting of violation suspected or actual.
	——	——

2. Checked transmission security. (Refer to ACP 124(D), ACP 125(E), ACP 126(C), FM 24-18, and unit SOI.)
    a. Ensured 20-second transmission time is being enforced.
    b. Ensured messages are preplanned before transmitting.
    c. Ensured equipment is operated in the secure mode.
	——	——

3. Checked cryptographic security utilizing the crypto facility checklist in TB 380-41.	——	——
4. Submitted completed results of the inspection to proper authorities.	——	——

Evaluation Guidance: Score the Soldier a GO if all PMs are passed. Score the Soldier a NO-GO if any PM is failed. If the Soldier fails any PM, show what was done wrong and how to do it correctly. Have the Soldier perform the PMs until they are done correctly.

References

Required	**Related**
ACP 124(D)	
ACP 125(E)	
ACP 126(C)	
AR 380-5	
FM 24-18	
FM 24-33	
TB 380-41	
UNIT SOI	

Inspect Installation of Antennas
113-596-7081

Conditions: Given the mast AB-155(*)/U (three each) or sufficient W-1 antenna wire to construct the doublet antenna, antenna group AN/GRA-50 (or equivalent) antenna system, magnetic compass, safety equipment, FM 24-18, TM 11-5820-467-15, TM 11-5985-357-13, and inspection checklist.

Standards: Conducted an inspection to ensure that the doublet antenna length was properly adjusted to frequency and was erected broadside to the distance station.

Performance Steps

1. Ensure antenna is installed properly using W-1 antenna wire.
    a. Ensure wire was cut to proper length for frequency.
    b. Ensure insulators were installed on both sides of antenna wire.
2. Inspect installation of Antenna Group AN/GRA-50 using Technical Manual. (Refer to TM 11-5820-467-15.)
3. Inspect azimuth settings.
4. Inspect guy line tension.
5. Inspect coaxial cable connections.
6. Inspect for the following safety connections.
    a. Guy cables rolled.
    b. Guys and stakes marked.
    c. Proper angle of guy stakes.

Performance Measures	GO	NO GO
1. Ensured antenna was properly installed using W-1 antenna wire.	——	——
2. Inspected installation of antenna using Antenna Group AN/GRA-50. (Refer to TM 11-5820-467-15.)	——	——
3. Inspected azimuth settings.	——	——
4. Inspected guy line tension.	——	——
5. Inspected coaxial cable connections.	——	——
6. Inspected safety precautions for guy cables and stakes.	——	——

Evaluation Guidance: Score the soldier GO if all steps are passed. Score the soldier NO-GO if any step is failed. If the soldier fails any step, show what was done wrong and how to do it correctly. Have the soldier practice until he can correctly perform the task.

References

Required	Related
FM 24-18	
TM 11-5820-467-15	
TM 11-5985-357-13	

STP 11-25C13-SM-TG

Inspect Installed Operational Radio Sets
113-620-7088

Conditions: Given an installed operational AN/GRC-106 (or equivalent) radio set with appropriate erected antenna, DA PAM 738-750, FM 24-18, TM 11-5820-520-10, inspection checklists, and DA Form 2404.

Standards: Inspected all components of the radio set and inspected the antenna for proper installation and grounding, corrected or reported all deficiencies, and submitted all reports and maintenance forms as required.

Performance Steps

1. Obtain appropriate references for the radio or radio set you will inspect.

2. Check siting of radio set. (Refer to FM 24-18.)
    a. Technical requirements: primary and backup sites.
    b. Tactical requirements: cover and concealment.
    c. Security requirements: physical and communications.

3. Check grounding system of radio set. (Refer to TM 11-5820-520-10.)

4. Check antenna system of radio set. (Refer to FM 24-18.)
    a. Location
    b. Installed in accordance with TM and meets all safety requirements.

5. Check equipment mountings. (Refer to TM 11-5820-520-10.)

6. Check tuned radio set. (Refer to TM 11-5820-520-10.)
    a. Check indicators frequently.
    b. Use operational and equipment performance checklist to correct faults.
    c. If correction cannot be made, report condition to Electronics Maintenance Shop.

7. Submit required reports. (Refer to DA PAM 738-750.)
    a. Meaconing, intrusion, jamming and interference reports.

8. Review maintenance forms for completeness and accuracy.

Performance Measures	GO	NO GO
1. Obtained appropriate references for the radio or radio set you inspected.	——	——
2. Checked siting of radio set. (Refer to FM 24-18.)	——	——
3. Checked grounding system of radio set. (Refer to TM 11-5820-520-10.)	——	——
4. Checked antenna system of radio set. (Refer to FM 24-18.)	——	——
5. Checked equipment mountings. (Refer to TM 11-5820-520-10.)	——	——
6. Checked tuned radio set. (Refer to TM 11-5820-520-10.)	——	——
7. Submitted required reports. (Refer to DA PAM 738-750.)	——	——
8. Reviewed maintenance forms for completeness and accuracy.	——	——

Evaluation Guidance: Score the soldier GO if all steps are passed. Score the soldier NO-GO if any step is failed. If the soldier fails any step, show what was done wrong and how to do it correctly. Have the soldier practice until he can correctly perform the task.

References

Required
DA FORM 2404
DA PAM 738-750
FM 24-18
TM 11-5820-520-10

Related
TM 11-5820-919-12
TM 11-5820-923-12
TM 11-5820-924-13

STP 11-25C13-SM-TG

Subject Area 11: GENERATORS AND EQUIPMENT PREVENTATIVE MAINTENANCE CHECKS AND SERVICES

Inspect Installed Operational Generator Sets
113-601-7055

Conditions: Given TM 5-6115-332-14, inspection checklist, an installed operational generator set 5 KW (or equivalent), and personnel.

Standards: Inspected the generator set for proper installation and grounding and met the required safety guidelines.

Performance Steps

1. Check generator set for proper grounding. (Refer to TM 5-6115-332-14.)
2. Check power cable for proper connections and wiring. (Refer to TM 5-6115-332-14.)
3. Ensure output voltage and frequency meet the requirements of the equipment being supplied power.
4. Ensure all canvas is rolled up or removed from the generator set.
5. Check fuel hoses for proper connections. (Refer to TM 5-6115-332-14.)
6. Check all meter readings for correct indications. (Refer to TM 5-6115-332-14.)

Performance Measures	GO	NO GO
1. Checked generator set for proper grounding. (Refer to TM 5-6115-332-14.)	——	——
2. Checked power cable for proper connections and wiring. (Refer to TM 5-6115-332-14.)	——	——
3. Ensured output voltage and frequency met the requirements of the equipment being supplied power.	——	——
4. Ensured all canvas was rolled up or removed from the generator set.	——	——
5. Checked fuel hoses for proper connections. (Refer to TM 5-6115-332-14.)	——	——
6. Checked all meter readings for correct indications. (Refer to TM 5-6115-332-14.)	——	——

Evaluation Guidance: Score the Soldier a GO if all PMs are passed. Score the Soldier a NO-GO if any PM is failed. If the Soldier fails any PM, show what was done wrong and how to do it correctly. Have the Soldier perform the PMs until they are done correctly.

References
Required
TM 5-6115-332-14

Related
TM 5-6115-271-14
TM 5-6115-275-14
TM 5-6115-365-15

STP 11-25C13-SM-TG

Inspect the Preventive Maintenance Checks and Services (PMCS)
113-617-7119

Conditions: As a communications systems supervisor, you are required to check the PMCS performed by assigned personnel. You are provided with assigned TOE equipment, assigned personnel, authorized parts and materials, AR 750-1, DA Form 2404, and DA PAM 738-750.

Standards: Inspected all required PMCS for completeness on maintenance forms.

Performance Steps

1. The cornerstone of unit maintenance is operator/crew performance of PMCS from the applicable TMs (-10 and –20 series). The BEFORE, DURING, and AFTER PMCS concentrate on ensuring equipment is fully mission capable.

2. To ensure adequate time is provided for performance of PMCS, you will coordinate with the training officer/NCO for scheduling maintenance time on the training schedule.

3. The performance of PMCS must be accomplished within the authorized maintenance level. You will monitor operator/crew performance of PMCS to ensure these are performed properly. Because of PMCS requirements, you will need to refer to the appropriate equipment TMs.

4. Coordinate with the unit motor sergeant for technical assistance in performing PMCS for vehicle and generator equipment, and with the electronic maintenance NCO for assistance for Signal equipment.

5. Once PMCS are completed, you must ensure DA Form 2404 is updated to reflect the date and the operator's initials are in block 10c. If uncorrected faults remain after PMCS are completed, these must be recorded on DA Form 2404 in block 10c, and the readiness status recorded in block 10b.

6. When PMCS reveal deficiencies requiring higher-level maintenance, you must ensure DA Form 2407 (Maintenance Request) is completed and the equipment is turned in for repair.

7. Equipment logbooks must be maintained on some equipment. You must ensure that logbook forms are properly completed and maintained. DA PAM 738-750 identifies the equipment requiring logbooks.

Evaluation Preparation: Setup: TOE assigned equipment, assigned personnel, appropriate equipment TMs, and authorized parts and materials will be available.

Brief Soldier: You will direct the performance of PMCS on assigned TOE equipment.

Performance Measures	GO	NO GO
1. Coordinated the scheduling of maintenance time and training schedules with the training officer/NCO.	——	——
2. Ensured required supplies, equipment, and technical publications were available and utilized.	——	——
3. Ensured equipment operators performed PMCS within their authorized level of maintenance as outlined in the applicable TMs.	——	——
4. Ensured correct maintenance procedures, as outlined in applicable TMs, were followed.	——	——

5. Coordinated with the applicable section for technical assistance.
   a. Motor sergeant for vehicle and generator equipment.
   b. Battalion electronic maintenance for Signal equipment.
	——	——

Performance Measures	<u>GO</u>	<u>NO GO</u>

6. Ensured DA Form 2404 reflected: ——   ——
   a. Inspection and services that were completed.
   b. Uncorrected faults.
   c. Readiness status.

7. Directed submission of DA Form 2407 as required.   ——   ——

8. Ensured equipment logbook and forms were completed and maintained IAW DA PAM 738-750.   ——   ——

9. Reported the readiness status of all equipment to the maintenance officer/NCO.   ——   ——

Evaluation Guidance: Score the Soldier a GO if all PMs are passed. Score the Soldier a NO-GO if any PM is failed. If the Soldier fails any PM, show what was done wrong and how to do it correctly. Have the Soldier perform the PMs until they are done correctly.

References

Required	**Related**
AR 750-1	
DA FORM 2404	
DA PAM 738-750	

This page intentionally left blank.

STP 11-25C13-SM-TG

CHAPTER 4

Duty Position Tasks

ADDITIONAL SKILL IDENTIFIER (ASI) T1

Subject Area 14: ENHANCED POSITION LOCATION REPORTING SYSTEM

Operate Generator Set MEP-803A
113-601-2056

Conditions: Given an MEP-803A generator set and TM 9-6115-642-10.

Standards: Checked, started, and stopped the generator set within 30 minutes.

Performance Steps

1. Perform all preoperational checks. (Refer to TM 9-6115-642-10, Chapter 2, Table 2-4.)
   a. Check generator set exterior.
   b. Check cooling system.
   c. Check exhaust/intake system.
   d. Check grounding rod assembly.
   e. Check electrical system.
   f. Check control box assembly.
   g. Check fuel level.

2. Install ground rod. (Refer to TM 9-6115-642-10, Chapter 2, Section III, para 2-8.1.)

DANGER: DO NOT operate the generator set until it is connected to a suitable ground. Serious injury or DEATH can result from operating an ungrounded generator set.

3. Install load cables. (Refer to TM 9-6115-642-10, Chapter 2, Section III, para 2-8.2.)

4. Start MEP-803A generator set. (Refer to TM 9-6115-642-10, Chapter 2.)

WARNING: Never attempt to start the generator set if it is not properly grounded. DO NOT crank engine in excess of 15 seconds. Allow starter to cool at least 15 seconds between attempted starts. Failure to observe this caution could result in damage to the starter.

   a. Rotate MASTER SWITCH to START position.
   b. Hold MASTER SWITCH in START position until oil pressure gauge reads at least 25 psi, voltage reads normal voltage requirements, and engine has reached a stable operating speed.
   c. Release MASTER SWITCH to PRIME AND RUN position.

NOTE: Under normal conditions warm up engine without load for 5 minutes.

   d. Check COOLENT TEMP (170-200 degree F) and OIL PRESSURE (25-60 psi) for normal readings.
   e. Using VOLTAGE Adjust potentiometer (refer to TM 9-6115-642-10, Chapter 2, Figure 2-1) and Frequency Adjust Control (refer to TM 9-6115-642-10, Chapter 2, Figure 2-3), adjust voltage and frequency to rated values.
   f. Press GROUND FAULT CIRCUIT INTERRUPTER TEST pushbutton. Ensure indicator window is clear. Press RESET pushbutton and ensure indicator is red.
   g. Ensure frequency and voltage are still at required values. Adjust if necessary.
   h. Rotate AM-VM transfer switch to each phase position while observing ammeter (PERCENT RATED CURRENT meter). If more than rated load is indicated in any phase, reduce load.

16 February 2005   4-1

Performance Steps

DANGER: High voltage is produced when this generator set is in operation. Improper operation could result in personal injury or DEATH.

CAUTION: With any access door open, the noise level of this generator set when operating could cause hearing damage. Hearing protection must be worn when working near the generator set while running.

    i. Perform all DURING (D) OPERATION PMCS requirements IAW TM 9-6115-642-10, Chapter 2, Table 2-4.

5. Stop MEP-803A generator set. (Refer to TM 9-6115-642-10, Chapter 2.)
   a. Place AC CIRCUIT INTERRUPTER switch in OPEN position.
   b. Allow generator set to operate 5 minutes with no load applied.
   c. Place MASTER SWITCH in OFF position.
   d. Perform all AFTER OPERATION (A) PMCS requirements IAW TM 9-6115-642-10, Chapter 2, Table 2-4.
   e. Place DEAD CRANK switch to OFF position.

NOTE: To perform Emergency Stopping Procedures, depress the EMERGENCY STOP pushbutton.

Evaluation Preparation: Setup: Ensure that all information, references, and equipment required to perform the task are available. Use the technical manual and the evaluation guide to score the Soldier's performance.

Brief Soldier: Tell the Soldier what the requirement is IAW the task conditions and standards.

Performance Measures	GO	NO GO
1. Performed all preoperational checks.	——	——
2. Installed ground rod.	——	——
3. Installed load cables.	——	——
4. Started MEP-803A generator set.	——	——
5. Stopped MEP-803A generator set.	——	——

Evaluation Guidance: Score the Soldier a GO if all PMs are passed. Score the Soldier a NO-GO if any PM is failed. If the Soldier fails any PM, show what was done wrong and how to do it correctly. Have the Soldier perform the PMs until they are done correctly.

References
    Required   **Related**
    TM 9-6115-642-10

STP 11-25C13-SM-TG

Prepare the EPLRS Net Control Station for Movement
113-630-1002

Conditions: Given EPLRS net control station (NCS), three-man team, MEP-803A generator set, TM 11-5825-295-10, and TM 9-6115-642-10.

Standards: Powered down the EPLRS net control station and secured equipment/cables to prevent movement.

Performance Steps
(Refer to TM 11-5825-295-10 for all performance steps, unless otherwise indicated.)

1. Perform system power down procedures. (Refer to TM 11-5825-295-10, Section IX.)

2. Ensure generator set was powered down. (Refer to TM 9-6115-642-10, Chapter 2.)

3. At POWER ENTRANCE PANEL on NCS, disconnect power cable from EXTERNAL POWER in receptacle. Replace dust covers.

4. Remove generator ground cable. (Refer to TM 9-6115-642-10.)
    a. Remove grounding system.
    b. Stow ground cable and grounding system.

5. Disconnect power cable from generator and stow in NCS storage.

6. Pull out EXTERNAL POWER circuit breaker on NCS POWER ENTRANCE PANEL.

7. Remove antennas.
    a. Remove DECRU antenna AS-3449 from antenna mast AB-1386/U.
    b. Secure antenna mast AB-1386/U and components.
    c. Remove SINCGARS antenna AS-3900.
    d. Remove NCS radio set antenna AS-3449.
    e. Stow antennas in NCS storage.

8. Disconnect shelter ground cable from SHELTER GROUND connector on POWER ENTRANCE PANEL.
    a. Remove ground system.
    b. Stow ground cable and grounding system in NCS storage.

9. Secure equipment within the NCS.

10. Stow first aid kit, fire extinguisher, and accessories.

11. Secure ECU covers.

12. Remove and stow ladder to shelter door.

13. Secure shelter door.

14. Connect power unit PU-798 to shelter mount for towing.

Evaluation Preparation: Setup: Ensure that all information, references, and equipment required to perform the task are available. Use the TMs and the evaluation guide to score the Soldier's performance.

Brief Soldier: Tell the Soldier what he is required to do IAW the task conditions and standards.

Performance Measures <u>GO</u> <u>NO GO</u>

1. Performed system power down procedures. ___ ___
2. Ensured generator set was powered down. ___ ___
3. Disconnected power cable from NCS and replaced dust covers. ___ ___
4. Removed generator ground cable. ___ ___
5. Disconnected power cable from generator and stowed in NCS storage. ___ ___
6. Pulled out EXTERNAL POWER circuit breaker on NCS POWER ENTRANCE PANEL. ___ ___
7. Removed antennas. ___ ___
8. Disconnected shelter ground cable. ___ ___
9. Secured equipment in the NCS. ___ ___
10. Stowed first aid kit, fire extinguisher, and accessories. ___ ___
11. Secured ECU covers. ___ ___
12. Removed and stowed ladder. ___ ___
13. Secured shelter door. ___ ___
14. Connected power unit PU-798 to shelter. ___ ___

Evaluation Guidance: Score the Soldier a GO if all PMs are passed. Score the Soldier a NO-GO if any PM is failed. If the Soldier fails any PM, show what was done wrong and how to do it correctly. Have the Soldier perform the PMs until they are done correctly.

References
   Required   **Related**
   TM 11-5825-295-10
   TM 9-6115-642-10

STP 11-25C13-SM-TG

Install the EPLRS Net Control Station
113-630-1005

Conditions: Given net control station (NCS), 10 kW generator set (PU-798), three-man team, TM 11-5825-295-10, TM 11-5985-426-12&P, and TM 9-6115-642-10.

Standards: Powered on the NCS, and the correct AC voltage readings were 100-120 volts and the DC voltage readings were 26-30 volts.

Performance Steps

1. Site the NCS and generator set on level area.

2. Emplace the NCS and generator set.
    a. Drive NCS to preselected level site and position vehicle so as to ensure antenna mast is no more than 10 degrees out of plumb when elevated.
    b. Position the generator set away from the NCS.

3. Ground the generator set. (Refer to TM 9-6115-642-10.)

4. Ground the NCS.

NOTE: Use protective eyewear and gloves when inserting ground rods and handling grounding cable.

   a. Assemble ground rod sections.
   b. Position ground rod within 10 feet of the NCS and drive into ground using sledgehammer.
   c. Connect one end of ground cable to ground rod with clamp.
   d. Connect other end of ground cable to SHELTER GROUND connector on shelter Power Entrance Panel.

NOTE: If using the MK-2551A/U, refer to TM 11-5825-295-10 for installation procedures.

5. Connect system power.
    a. Connect NATO plug to the DC connector panel located in the vehicle crew cab on the passenger side.
    b. Install ladder.
    c. Inside shelter, ensure that the DC MAIN circuit breaker is in the ON position.
    d. On the SHELTER LIGHTING panel, place DC LIGHTS in the ON position.
    e. Ensure that the AC MAIN breaker and all equipment switches are OFF.
    f. Connect power cable to EXTERNALPOWER IN connector at Power Entrance Panel.
    g. Connect power cable to generator set.

6. Install antennas.
    a. DECRU antenna on Mast AB-1386/U.
    b. NCS-RS and FM radio antennas.

7. Turn on prime power at generator. (Refer to TM 9-6115-642-10.)

8. Turn on power in the NCS ensuring that the AC VOLTS meter reads 100-120 volts and the DC VOLTS meter reads 26-30 volts.

NOTE: If readings are out of tolerance, refer to power source technical manual.

Evaluation Preparation: Setup: Ensure that all information, references, and equipment required to perform the task are available. Use the TMs and the evaluation guide to score the Soldier's performance.

Brief Soldier: Tell the Soldier what he is required to do IAW the task conditions and standards.

Performance Measures   <u>GO</u>   <u>NO GO</u>

1. Sited the NCS and generator set on level area.   ——   ——
2. Emplaced the NCS and generator set.   ——   ——
3. Grounded the generator set.   ——   ——
4. Grounded the NCS.   ——   ——
5. Connected system power.   ——   ——
6. Installed antennas.   ——   ——
7. Turned generator on.   ——   ——
8. Turned the NCS power on and checked voltages.   ——   ——

Evaluation Guidance: Score the Soldier a GO if all PMs are passed. Score the Soldier a NO-GO if any PM is failed. If the Soldier fails any PM, show what was done wrong and how to do it correctly. Have the Soldier perform the PMs until they are done correctly.

References
    Required   **Related**
    TM 11-5825-295-10
    TM 11-5985-426-12&P
    TM 9-6115-642-10

STP 11-25C13-SM-TG

Deploy the EPLRS Gateway Radio Set
113-630-1006

Conditions: Given one additional personnel; Radio Set AN/VSQ-2C(V)4; High mobility, multipurpose wheeled vehicle (HMMWV) M-998 with Installation Kit MK-2753/VSQ; Global Positioning System (GPS) Precision Lightweight GPS Receiver (PLGR) AN/PSN-11 or map and compass; Radio Set AN/VRC-90A (SINCGARS); TB 11-5825-283-23; TM 11-5985-426-12&P; TM 11-5820-890-10-1; TM 11-5825-291-13; and TM 11-5825-283-10.

Standards: The EPLRS Gateway Radio Set AN/VSQ-2C(V)4 passed the self-test and a message was sent and received between the division radio sets.

Performance Steps

1. Check equipment for completeness.

2. Install Radio Set AN/VSQ-2(V)4 in the HMMWV.
    a. Slide the selectable power adapter (SPA) and the RT-1720B(C)/G or RT-1720C(C)/G into the resilient mount MT-6146/VSQ-1 as a unit until two pins at rear of resilient mount engage in rear of SPA.
    b. Lift two retaining latches so that they engage the RT-1720B(C)/G or RT-1720C(C)/G and turn two knurled knobs clockwise to secure the unit.
    c. Secure ground leads to SPA and RT-1720B(C)/G or RT-1720C(C)/G respectively.
    d. Repeat steps a through c for the other SPA and RT.

3. Connect radio set cables.
    a. Connect URO extension cable connector P1 to Data J3 connector on radio set. Connect URO cable connector P1 to URO bracket connector on Dashboard Panel.
    b. Connect antenna cable connector P1 to ANT J2 connector on the radio set.
    c. Connect power cable connector P1 to J1 on the SPA.
    d. Repeat steps a through c for the other radio set.

4. Install antennas.
    a. Connect Antenna AS-3449/VSQ-1 to antenna bracket connector on left rear of the HMMWV.
    b. Connect Antenna AS-3449/VSQ-1 to antenna pedestal bracket connector of right rear of the HMMWV or connect antenna cable from Antenna AB-1386/U.

NOTE: If using Antenna Mast AB-1386/U, assemble the mast and antenna IAW TM 11-5985-426-12&P.

5. Site equipment.
    a. Use AN/PSN-11 to site equipment.
    b. If necessary, use compass.

6. Check voltage.
    a. On dashboard panel, set panel meter switch to on (up position).
    b. Check battery for proper voltage by turning meter switch on.

NOTE: If panel meter reads 22 volts or less, radio batteries require recharging.

7. Perform starting procedures.

NOTE: Ensure that the SINCGARS is installed and antenna connected within the HMMWV.

    a. Power on SINCGARS AN/VRC-90D.

NOTE: Refer to TM 11-5820-890-10-1 for SINCGARS operating procedures.

    (1) Load frequencies.
    (2) Load COMSEC keys.

STP 11-25C13-SM-TG

Performance Steps

      (3) Establish voice communications with Net Control Station.
   b. Power on Radio Set AN/VSQ-2(V)4.

NOTE: Refer to TM 11-5825-283-10 for power-up procedures.

      (1) Observe self-test.
      (2) Load COMSEC keys
         (a) Load different keys for different Divisions.

NOTE: Observe OUT OF NET indicator. If EPLRS network is available, indicator will stop flashing within one minute to show radio has entered the network.

  8. Enter cable length and link interface parameters.

  9. Connect the gateway cable to the J2 host interface connectors on the radio sets.

 10. Perform self-test on each radio set.

NOTE: Verify that TEST OK and same level of classification is indicated on radio sets.

 11. Perform a communications check between the radios by sending and receiving a message to ensure that the link is established between the two divisions.

Evaluation Preparation: Setup: Ensure that all information, references, and equipment required to perform the task are available. Use the TM and the evaluation guide to score the Soldier's performance.

Brief Soldier: Tell the Soldier what he is required to do IAW the task conditions and standards.

Performance Measures	GO	NO GO
1. Checked equipment.	——	——
2. Installed EPLRS radio sets.	——	——
3. Made cable connections.	——	——
4. Installed antennas.	——	——
5. Site equipment.	——	——
6. Checked for correct voltage readings.	——	——
7. Performed starting procedures.	——	——
8. Entered cable length and link interface parameters.	——	——
9. Connected gateway cable.	——	——
10. Performed self-test on radio sets.	——	——
11. Performed communications check by sending and receiving a message.	——	——

Evaluation Guidance: Score the Soldier a GO if all PMs are passed. Score the Soldier a NO-GO if any PM is failed. If the Soldier fails any PM, show what was done wrong and how to do it correctly. Have the Soldier perform the PMs until they are done correctly.

References
   Required   **Related**
   TB 11-5825-283-23
   TM 11-5820-890-10-1
   TM 11-5825-283-10

References
   Required　　　　　　　　　　　　　　　**Related**
    TM 11-5825-291-13
    TM 11-5985-426-12&P

STP 11-25C13-SM-TG

Deploy the Grid Reference Radio Set
113-630-1007

Conditions: Given one additional personnel, AN/GRC-229 Grid Reference Radio Set (GR-RS), AN/PSN-11 Precision Lightweight GPS Receiver (PLGR) or map and compass, TM 11-5825-283-10, and TM 11-5985-426-12&P.

Standards: Made a successful communications check with the net control station (NCS) and sent back a position (grid coordinates) of the radio set to the NCS.

Performance Steps

1. Check equipment for completeness.

2. Site equipment.

3. Assemble grid reference radio set (GR-RS).
   a. Install batteries in the GR-RS battery case.
   b. Slide receiver-transmitter (RT) and selectable power adapter (SPA) together.
   c. Engage RT latches with slots in SPA and turn butterfly handles clockwise to secure RT to SPA.
   d. Slide RT and SPA into GR-RS mount as a unit.
   e. Lift the two retaining latches so that they engage RT and turn two knurled knobs clockwise to secure RT to mount.
   f. Secure ground leads to RT and SPA.
   g. Connect RT to power source.
   h. Connect URO cable to DATA connector on RT.

4. Assemble antenna.
   a. Erect antenna IAW TM 11-5985-426-12&P.
   b. Secure EPLRS antenna adapter to antenna mast.
   c. Connect antenna cable to RF elbow connector and attach RF elbow to EPLRS antenna adapter.
   d. Secure AS-3449/VSQ-1 antenna element to EPLRS antenna adapter.

NOTE: Before raising the mast, check the antenna cable for cable length marking (in meters). If cable is not marked, estimate its length as closely as possible.

   e. Raise and secure AB-1386/U antenna mast.
   f. Connect antenna cable connector P1 to ANT connector on RT.

5. Power up the GR-RS.
   a. Set POWER switch to ON or ON+AUDIBLE.
   b. Verify self-test on radio set is complete.
   c. Verify that radio set displays @0.

NOTE: If radio set does not display @0, zeroize radio set before loading keys.

6. Load keys into the radio set. (Refer to TM 11-5825-283-10.)

7. Enter radio set antenna cable length.

NOTE: Use the cable length that you obtained in step 4d.

8. Perform communications check with the NCS and send back the position (grid coordinates) of the radio set to the NCS.

Evaluation Preparation: Setup: Ensure that all information, references, and equipment required to perform the task are available. Use the TMs and the evaluation guide to score the Soldier's performance.

Brief Soldier: Tell the Soldier what he is required to do IAW the task conditions and standards.

Performance Measures	GO	NO GO
1. Checked equipment for completeness.	——	——
2. Sited equipment.	——	——
3. Assembled the GR-RS.	——	——
4. Assembled antenna.	——	——
5. Powered up the GR-RS.	——	——
6. Loaded keys.	——	——
7. Entered cable length.	——	——
8. Performed communications check with the NCS and sent back the position (grid coordinates) of the radio set to the NCS.	——	——

Evaluation Guidance: Score the Soldier a GO if all PMs are passed. Score the Soldier a NO-GO if any PM is failed. If the Soldier fails any PM, show what was done wrong and how to do it correctly. Have the Soldier perform the PMs until they are done correctly.

References
    Required   **Related**
    TM 11-5825-283-10
    TM 11-5985-426-12&P

STP 11-25C13-SM-TG

Operate the EPLRS Net Control Station
113-630-2001

Conditions: Given an EPLRS net control station (NCS), COMSEC paper tapes, KOI-18 tape reader, AN/CYZ-10 data transfer device, EPLRS radio sets (community), installed 10 kW generator set, TM 11-5825-283-10, and TM 11-5825-295-10.

Standards: The EPLRS NCS came on-line and obtained a community.

Performance Steps
(Refer to TM 11-5825-295-10 for all performance steps.)

1. Perform system power-up procedures.
    a. Ensure that the DC main circuit breaker is in the ON position.
    b. On the SHELTER LIGHTING panel, place DC LIGHTS in the ON position.
    c. Ensure that the AC MAIN breaker and all equipment switches are off.
    d. Turn on prime power at generator.
    e. Push in EXTERNAL POWER breaker on power entrance panel.
    f. Place AC KILL switch in the AC position.
    g. Push in AC MAIN circuit breaker.
    h. Turn on both UPS units.
    i. Check that the AC VOLTSMETER reads 110-120 volts and the DC VOLTSMETER reads 26-30 volts.
    j. On the SHELTER LIGHTING panel, push the AC LIGHTS breaker in to turn on AC lights. Pull the DC LIGHTS breaker out to turn off DC LIGHTS.
    k. Turn on each environmental control unit.
    l. Turn on the KOK-13.
    m. Set DECRU POWER switch to ON position.
    n. On the NCS-RS, set POWER switch to ON+AUDIBLE.
    o. Power up data processing subsystem (TAC-4).

2. Perform operational key load procedures.
    a. Load seed keys into the KOK-13.
    b. Load initial keys into the AN/CYZ-10.
    c. Load keys into the DECRU.
    d. Load keys into the NCS-RS.

3. Perform system initialization procedures.
    a. Enter console login and password.
    b. Select F for live system operation.
    c. Load the RTEP program.
    d. Load/create library.
    e. Load/create map.

4. Place the System On-Line as Timing Master or as Non-Timing Master.

NOTE: If Timing Master, perform steps 5 through 7 on page 2-110 and additional steps on pages 2-112 through 2-114 in TM 11-5825-295-10. If Non-Timing Master, skip steps 5 through 7 on page 2-110 and additional steps on pages 2-114 through 2-116.

   a. Input system parameters.
   b. Enter any required changes to system. (Only if directed by NCOIC.)
      (1) Library changes.
      (2) Notices.
      (3) TREE allocation.

5. Perform system configuration procedures.

Performance Steps

6. Perform the following operational procedures.
   a. Ensure all full reference radio sets are on-line and are reporting correct position locations.
   b. Perform procedures to verify active needlines. Run a printout of these needlines.
   c. Print out all units within the library.
   d. Perform procedures to review/evaluate the following advisories.
      (1) System alert advisory.
      (2) Unit fault advisory.
      (3) No rekey advisory.
      (4) Zone penetration advisory.
      (5) Message advisory. (Perform only user message advisory and needline request advisory.)
      (6) No track advisory.
      (7) Unsatisfied needline advisory.
   e. Update unit and needline libraries.
   f. Update map.
   g. Record all library entries.
   h. Record map. Send a dispatch to another NCS.
   i. Perform a weekly rekey network request.
   j. Send a dispatch to another NCS.
   k. Set unit inactive.
   l. Restart that unit.

7. Perform data logging procedures.

8. Ensure all radio sets (units) are in the net.

Evaluation Preparation: Setup: Ensure that all information, references, and equipment required to perform the task are available. Use the TM and the evaluation guide to score the Soldier's performance.

Brief Soldier: Tell the Soldier what he is required to do IAW the task conditions and standards.

Performance Measures	GO	NO GO
1. Performed system power-up procedures.	——	——
2. Performed operational key load procedures.	——	——
3. Performed system initialization procedures.	——	——
4. Placed the system on-line as Timing Master or Non-Timing Master.	——	——
5. Performed system configuration procedures.	——	——
6. Performed operational procedures.	——	——
7. Performed data logging procedures.	——	——
8. Ensured all radio sets (units) were in the net.	——	——

Evaluation Guidance: Score the Soldier a GO if all PMs are passed. Score the Soldier a NO-GO if any PM is failed. If the Soldier fails any PM, show what was done wrong and how to do it correctly. Have the Soldier perform the PMs until they are done correctly.

References

Required	Related
TM 11-5825-283-10	
TM 11-5825-295-10	

STP 11-25C13-SM-TG

Perform ECCM Procedures
113-630-2027

Conditions: Given an EPLRS net control station and TM 11-5825-295-10.

Standards: The EPLRS net control station overcame jamming.

Performance Steps

1. Advise supervisor.

2. Perform frequency hop/no hop request.
   a. Set frequency to hop mode.

3. Perform power level change.
   a. Set power level to MED HI.

4. Identify radio sets that are most affected by jamming and increase their report rates.

5. Reestablish a stable EPLRS community when jamming has subsided or stopped.

Evaluation Preparation: Setup: Ensure that all information, references, and equipment required to perform the task are available. Use the TM and the evaluation guide to score the Soldier's performance.

Brief Soldier. Tell the Soldier what the requirement is IAW the task conditions and standards.

Performance Measures	GO	NO GO
1. Advised supervisor.	——	——
2. Performed frequency hop/no hop request.	——	——
3. Performed power level change.	——	——
4. Identified radio sets and increased their report rates.	——	——
5. Reestablished a stable EPLRS community.	——	——

Evaluation Guidance: Score the Soldier a GO if all PMs are passed. Score the Soldier a NO-GO if any PM is failed. If the Soldier fails any PM, show what was done wrong and how to do it correctly. Have the Soldier perform the PMs until they are done correctly.

References
    Required   **Related**
    TM 11-5825-295-10

STP 11-25C13-SM-TG

Perform Playback Mode Operational Procedures
113-630-2031

Conditions: Given an operational Net Control Station AN/TSQ-158(V)4, data logs/tapes, and TM 11-5825-295-10.

Standards: The supervisor could receive data of a previous live operation/exercise.

Performance Steps

(Refer to TM 11-5825-295-10 for all performance steps.)

1. Perform playback mode load/initialization procedure.
   a. Load appropriate playback files.
   b. Start playback.

2. Perform playback rate selection.

3. Perform start time selection.

4. Perform print times selection.

Evaluation Preparation: Setup: Ensure that all information, references, and equipment required to perform the task are available. Use the TM and the evaluation guide to score the Soldier's performance.

Brief Soldier. Tell the Soldier what he is required to do IAW the task conditions and standards.

Performance Measures	**GO**	**NO GO**
(Refer to TM 11-5825-295-10 for all PMs.)		
1. Performed playback mode load/initialization procedure.	——	——
2. Performed playback rate selection.	——	——
3. Performed start time selection.	——	——
4. Performed print times selection.	——	——

Evaluation Guidance: Score the Soldier a GO if all PMs are passed. Score the Soldier a NO-GO if any PM is failed. If the Soldier fails any PM, show what was done wrong and how to do it correctly. Have the Soldier perform the PMs until they are done correctly.

References
    Required   **Related**
    TM 11-5825-295-10

STP 11-25C13-SM-TG

Operate the EPLRS Radio Set
113-630-2044

Conditions: This task is performed in an EPLRS network. Soldier will be given Data Transfer Device AN/CYZ-10 with fill cable; EPLRS Radio Set; two BA-5590/U batteries or two BB-390A batteries; selectable power adapter; TM 11-5825-283-10, and TM 11-5825-299-10.

Standards: The radio set entered into net and could send and receive messages.

Performance Steps

1. Prepare radio set for operation.
   a. Install selectable power adapter (SPA) onto the radio set.
   b. Connect P1 of power cable into J1 on right side of power adapter.
   c. Connect P1 of antenna cable into ANT connector J2 on RT front panel.
   d. Connect P1 of URO extension cable into DATA connector J3 on RT front panel.
   e. Connect P1 of URO cable to connector P2 of URO extension cable.
   f. On right side of SPA, press in circuit breaker CB1.

2. Obtain radio set parameters checklist from the NCOIC.

3. At the radio set, set POWER switch to ON or ON+AUDIBLE.
   a. Observe self-test results to ensure URO displays an @0 and radio set has no faults.

4. Send a message and ensure that the guard channel and radio set ID are correct.

5. Send a Q message to verify the software and hardware versions against the radio set parameters checklist.

6. Load keys into the radio set.

7. Observe that the radio set's out-of-net indicator goes out (indicating radio is in net).

8. Send a C message to enter the radio set antenna cable length.

9. Send an A message to enter the difference in height between the RT and the antenna.

10. Check DIVID against the radio set parameters checklist.

NOTE: If DIVID needs to be changed, send a 'degree a' message. If operating with the net control station, perform step 11. If operating with the EPLRS network manager (ENM), perform step 12.

11. Send a freetext message to the net control station. (Refer to TM 11-5825-283-10.)

12. Send an S message to all radio sets. (Refer to TM 11-5825-299-10.)

13. Display incoming message by pressing the receive key on URO.

Performance Measures	**GO**	**NO GO**
1. Prepared radio set for operation.	——	——
2. Obtained radio set parameters checklist.	——	——
3. Performed power-up procedures.	——	——
4. Sent a message to ensure guard channel and radio set ID were correct.	——	——
5. Sent a Q message and verified software and hardware versions were correct.	——	——
6. Loaded keys into the radio set.	——	——

Performance Measures	<u>GO</u>	<u>NO GO</u>
7. Ensured radio set was in the net.	——	——
8. Sent a C message and verified/corrected antenna cable length.	——	——
9. Sent an A message and entered the difference in height between the RT and the antenna.	——	——
10. Checked DIVID.	——	——
11. If operating with the net control station, sent a freetext message.	——	——
12. If operating with the ENM, sent an S message.	——	——
13. Received a message.	——	——

Evaluation Guidance: Score the Soldier a GO if all PMs are passed. Score the Soldier a NO-GO if any PM is failed. If the Soldier fails any PM, show what was done wrong and how to do it correctly. Have the Soldier perform the PMs until they are done correctly.

References

    Required　　　　　　　　　　　　　　　　　**Related**
    TM 11-5825-283-10　　　　　　　　　　　　TM 11-5825-283-10-1
    TM 11-5825-299-10

STP 11-25C13-SM-TG

Perform Multiple Net Control Station (NCS) Community Operational Procedures
113-630-2045

Conditions: The net control station (NCS) has been notified to assume control of the network. Given an EPLRS NCS that has been powered up with all required hardware and software plus ancillary equipment needed to provide support, EPLRS radio sets in an established NCS-controlled community with references, installed 10 KW Generator Set, TM 11-5825-283-10, and TM 11-5825-295-10. Individual AOR will be used for both NCSs and a planned handover will occur to split community.

Standards: The NCS established an AOR in a multiple NCS community, assumed control of references and units (radio sets), conducted a successful planned handover, and cleared all advisories.

Performance Steps

(Refer to TM 11-5825-295-10 for all performance steps.)

1. Enter adjacent NCS report request.
   a. Hook NCS.
   b. Monitor unit link tabular display area.

2. Request adjacent NCS data/time.

3. Suggest secondary NCS assignment.
   a. Determine intercommunity contacts.
      (1) Hook NCS.
      (2) Monitor unit link tabular display area.
   b. Hook a unit that has good communications with the NCS and perform the steps to make that unit the secondary NCS assignment. (Refer to TM 11-5825-295-10, page 2-183.)

4. Perform planned handover.
   a. Notify gaining NCS of planned shutdown.
   b. Make all radio sets handover eligible.
   c. Coordinate AOR expansion and reduction.
   d. Conduct manual radio set handover of at least three radio sets.
   e. Gaining NCS confirms acquisition of at least three radio sets.

5. Monitor and resolve advisories.
   a. Alert advisory.
   b. Unit Fault advisory.
   c. No Track advisory.
   d. Unset Needline advisory.
   e. Reference Quality advisory.

Evaluation Preparation: Setup: Ensure that all information, references, and equipment required to perform the task are available. Use the TMs and the evaluation guide to score the Soldier's performance.

Brief Soldier. Tell the Soldier what he is required to do IAW the task conditions and standard.

Performance Measures	GO	NO GO
1. Entered adjacent NCS report request.	——	——
2. Requested adjacent NCS data/time.	——	——
3. Suggested secondary NCS assignment.	——	——
4. Performed planned handover.	——	——
5. Monitored and resolved advisories.	——	——

STP 11-25C13-SM-TG

Evaluation Guidance: Score the Soldier a GO if all PMs are passed. Score the Soldier a NO-GO if any PM is failed. If the Soldier fails any PM, show what was done wrong and how to do it correctly. Have the Soldier perform the PMs until they are done correctly.

References
    Required   **Related**
    TM 11-5825-283-10
    TM 11-5825-295-10

STP 11-25C13-SM-TG

Perform Preventive Maintenance Checks and Services (PMCS) on the EPLRS Net Control Station
113-630-3011

Conditions: Given an EPLRS net control station (NCS) needing PMCS, Tool Kit TK-17/G or equivalent, cotton lint-free cloth, typewriter brush, isopropyl alcohol, cleaning cartridge, TM 11-5825-295-10, TM 10-5411-222-14, TM 11-5805-780-15, TM 11-5820-890-10-1, TM 11-5825-283-10, TM 11-5825-291-13, DA PAM 738-750, and DD Form 314 (Preventive Maintenance Schedule and Record).

Standards: Completed all PMCS and identified equipment as ready/available or not ready/available.

Performance Steps

1. Review maintenance forms.

2. Perform PMCS routine services.

3. Perform PMCS at the prescribed interval.
    a. Check NCS-RS keep-alive battery.
    b. Check DECRU keep-alive battery.
    c. Check the battery for AN/CYZ-10.
    d. Check the KOI-18 battery.
    e. Clean Data Processing Subsystem (DPS) TAC-4 Digital Tape Drive Cleaning.
        (1) Insert cleaning cartridge into tape drive. (Tape drive loads cartridge, cleans heads, and ejects cartridge.)
        (2) Remove cartridge and log date that the drive was cleaned on cartridge label.
    f. Check the KOK-13 Internal keep-alive battery.

NOTE: It is extremely important that the KOK-13 internal keep-alive battery check is done daily.

4. Complete all required maintenance forms.

5. Schedule next PMCS.

Evaluation Preparation: Setup. Ensure that all information, references, and equipment required to perform the task are available. Use the TMs and the evaluation guide to score the Soldier's performance.

Brief Soldier. Tell the Soldier what is required to do IAW the task conditions and standards.

Performance Measures	**GO**	**NO GO**
1. Reviewed maintenance forms.	——	——
2. Performed PMCS routine services.	——	——
3. Performed PMCS at the prescribed interval.	——	——
4. Completed all required maintenance forms.	——	——
5. Scheduled next PMCS on DD Form 314.	——	——

Evaluation Guidance: Score the Soldier a GO if all PMs are passed. Score the Soldier a NO-GO if any PM is failed. If the Soldier fails any PM, show what was done wrong and how to do it correctly. Have the Soldier perform the PMs until they are done correctly.

References
  Required
  DA PAM 738-750
  DD FORM 314
  TM 10-5411-222-14
  TM 11-5820-890-10-1
  TM 11-5825-283-10
  TM 11-5825-291-13
  TM 11-5825-295-10

  Related
  TM 11-5815-602-24
  TM 11-5825-272-23
  TM 5-4120-379-14
  TM 5-4120-384-14

STP 11-25C13-SM-TG

Troubleshoot the EPLRS Net Control Station
113-630-3014

Conditions: Given a faulty net control station (NCS), tool kit TK-17, multimeter AN/PSM-45A, TM 11-5825-295-10, and TM 11-5825-295-23&P.

Standards: The NCS successfully communicated with the EPLRS community.

Performance Steps

1. Perform system power-up procedures. (Refer to TM 11-5825-295-10, Chapter 2, Section III.) If any fault occurs, go to performance step 6.
    a. On the Shelter Management Panel, place the DC MAIN circuit breaker in the ON position (pushed in).
    b. On the Shelter Management Panel, place DC LIGHTS in the ON position.
    c. On the Shelter Management Panel, observe the EXTERNAL POWER light is lit, indicating that external power is applied to the shelter.
    d. On the Shelter Management Panel, place AC/KILL switch to the AC position.
    e. On the Shelter Management Panel, push in the AC MAIN circuit breaker.
    f. Turn ON the Uninterruptible Power Supplies (UPS).
    g. Check that the AC Volts meter reads 110-120 volts and the DC Volts meter reads 26-30 volts. If readings are out of tolerance, refer to power source technical manual for procedures on corrective adjustments.
    h. On the Shelter Management Panel, turn the AC LIGHTS on and the DC LIGHTS off.

2. Perform equipment power-up procedures. If any fault occurs, go to performance step 6.
    a. Turn on each Environmental Control Unit (ECU) as required.
    b. Turn on KOK-13.
    c. Set DECRU POWER switch to ON position.
    d. On NCS-RS, set POWER switch to ON+AUDIBLE.
    e. Power up Data Processing Set (DPS) (TAC-4) and associated equipment.

3. Perform operational key load procedure. (Refer to TM 11-5825-295-10, Chapter 2, Section III.)

4. Perform system initialization procedures. (Refer to TM 11-5825-295-10, Chapter 2, Section III.) If fault or alert message occurs, verify it is not an operator correctable fault; otherwise go to performance step 6.

NOTE: During system initialization procedures, if RTEP program (MATH DATA) does not load correctly, go to performance step 6.

5. Place the NCS on-line. (Refer to TM 11-5825-295-10, Chapter 2, Section III.) If faults or alert message occurs, go to performance step 6.

6. Identify the fault symptoms. (Refer to TM 11-5825-295-23&P, Chapter 2.)

7. Perform performance steps 8 through 12. (Refer to TM 11-5825-295-23&P, Chapter 2, Section V.)

8. Perform System Level Diagnostic Program (SLDP) procedures, if necessary. (Refer to Chapter 2, Section V of.) If fault is detected, go to performance step 12.

9. Perform secondary DECRU/DPS Input/Output Loopback Test, if necessary. (Refer to TM 11-5825-295-23&P, Chapter 2, Section V.) If fault is detected, go to performance step 12.

10. Perform NCS/DPS Input/Output Loopback Test, if necessary. (Refer to TM 11-5825-295-23&P, Chapter 2, Section V.) If fault is detected, go to performance step 12.

11. Perform X-Windows System Tool Manager (XSTM) procedure, if necessary. (Refer to TM 11-5825-295-23&P, Chapter 2, Section V.) If fault is detected, go to performance step 12.

STP 11-25C13-SM-TG

Performance Steps

12. When the fault is identified, select the correct fault isolation procedure.

13. Isolate to the Shop Replaceable Unit (SRU).

14. Perform corrective action. Go to performance step 15.

15. Remove faulty SRU. (Refer to TM 11-5825-295-23&P, Chapter 2, Section V.)

16. Replace faulty SRU, if necessary.

NOTE: If SRU replacement is not required, go to performance step 17.

17. Verify the fault is fixed, perform required after maintenance checks.

18. Isolate to the Line Replaceable Unit (LRU).

19. Perform corrective action. Go to performance step 20.

20. Remove faulty LRU. (Refer to TM 11-5825-295-23&P, Chapter 2, Section V.)

21. Replace faulty LRU, if necessary.

NOTE: If LRU replacement is not required, go to performance step 22.

22. Verify the fault is fixed, perform required after maintenance checks.

23. Repeat performance steps 1 through 5. If no other faults are found, go to performance step 24.

24. Resume operations. (Refer to TM 11-5825-295-10.)

Performance Measures	**GO**	**NO GO**
1. Performed system power-up.	——	——
2. Checked AC voltmeter and DC voltmeter for correct readings.	——	——
3. Set remaining power control panel circuit breakers to ON.	——	——
4. Performed power-up procedures.	——	——
5. Performed operational key load procedure.	——	——
6. Performed RTEP program load procedure.	——	——
7. Performed system initialization procedures.	——	——
8. Placed the NCS on-line.	——	——
9. Identified fault symptoms.	——	——
10. Performed PMs 8 through 11, as applicable.	——	——
11. Performed System Level Diagnostic Program (SLDP), if necessary.	——	——
12. Performed secondary DECRU/DPS/Input/Output Loopback Test, if necessary.	——	——
13. Performed NCS/DPS/ Input/Output Loopback Test, if necessary.	——	——
14. Performed X-Windows System Tool Manager (XSTM) procedure, if necessary.	——	——
15. When fault was identified, selected the correct fault isolation procedure.	——	——
16. Performed corrective action.	——	——

16 February 2005

Performance Measures	<u>GO</u>	<u>NO GO</u>
17. Removed and replaced faulty SRU.	——	——
18. Verified that the fault was fixed (performed required after maintenance checks).	——	——
19. Isolated to the Line Replaceable Unit (LRU).	——	——
20. Performed corrective action.	——	——
21. Removed and replaced faulty LRU.	——	——
22. Verified that the fault was fixed (performed required after maintenance checks).	——	——
23. Repeated PMs 1 through 8. If no other fault was found, went to PM 24.	——	——
24. Resumed operations.	——	——

Evaluation Guidance: Score the Soldier a GO if all PMs are passed. Score the Soldier a NO-GO if any PM is failed. If the Soldier fails any PM, show what was done wrong and how to do it correctly. Have the Soldier perform the PMs until they are done correctly.

References
    Required   **Related**
    TM 11-5825-295-10
    TM 11-5825-295-23&P

APPENDIX A – DA Form 5164-R (Hands-On Evaluation)

A-1. Introduction.

The DA Form 5164-R allows the trainer to keep a record of the performance measures a Soldier passes or fails on each task. Instructions for using this form follow.

A-2. Prior to evaluating the Soldier.

a. Obtain a blank copy of DA Form 5164-R, which you may locally reproduce on 8 1/2- by 11-inch paper.

b. Enter the task title and 10-digit task number from the Soldier's manual task summary in Chapter 3.

c. In column a, enter the number of each PM listed under the Performance Measures section in the task summary.

d. In column b, enter the PM corresponding to the PM number in column a. (You may abbreviate this information if necessary.)

e. Enter the Evaluation Guidance statement from the Soldier's manual task summary just below the last PM.

f. Locally reproduce the partially completed form if you are evaluating more than one Soldier on the task or the same Soldier on more than one task.

A-3. During the evaluation.

a. Enter the date just before evaluating the Soldier's task performance.

b. Enter the evaluator's name and the Soldier's name and unit.

c. For each PM, column b, enter a check in column c *(PASS)* or column d *(FAIL)*, as appropriate.

d. Compare the number of PMs the Soldier passes (and if applicable, which ones) against the task standard shown in the Evaluation Guidance statement. If the standard is met or exceeded, check the *GO* block under *STATUS*; otherwise check the *NO-GO* block.

Figure A-1 is a sample of a completed DA Form 5164-R.

STP 11-25C13-SM-TG

HANDS-ON EVALUATION		DATE
For use of this form, see AR 350-57. The proponent agency is DCSOPS.		2 FEB 2004
TASK TITLE: PERFORM SYSTEM SHUTDOWN FOR SEN SYSTEM AN/TTC-48(V)		TASK NUMBER: 113-625-2090

ITEM a	PERFORMANCE STEP b	SCORE (Check One)	
		PASS c	FAIL d
1	PERFORMED OPERATIONAL SHUTDOWN PROCEDURES	✓ P	F
2	PERFORMED STORAGE PROCEDURES	✓ P	F
3	PERFORMED POWER CABLING REMOVAL/STORAGE PROCEDURES	✓ P	F
4	PERFORMED SUBSCRIBER FIELD CABLE REMOVAL/STORAGE PROCEDURES	✓ P	F
5	PERFORMED GROUNDED STRAP AND ROD REMOVAL PROCEDURES	P	✓ F
6	SECURED THE SHELTER DOOR AND ALL EXTERNAL COVERS	✓ P	F

EVALUATOR'S NAME: SFC WHITMAN	UNIT: A CO 369TH
SOLDIER'S NAME: SPC ANDERSON	STATUS: ☐ GO ☑ NO GO

DA FORM 5164-R, SEP 85 (EDITION OF DEC 82 IS OBSOLETE)

Figure A-1. Sample DA Form 5164-R

STP 11-25C13-SM-TG

APPENDIX B - DA Form 5165-R (Field Expedient Book)

B-1. Introduction.

The DA Form 5165-R allows the trainer to keep a record of task proficiency for a group of Soldiers. Instructions for using this form follow.

B-2. Prior to evaluating the Soldier.

a. Obtain a blank copy of DA Form 5165-R, which you may locally reproduce on 8 1/2- by 11-inch paper.

b. Enter the SM task number and abbreviated task title for the evaluated tasks in the appropriate column. Use additional sheets as necessary.

c. Locally reproduce the partially completed form if you are evaluating more than nine Soldiers.

B-3. During the evaluation.

a. Enter the names of the Soldiers you are evaluating, one name per column, at the top of the form. You may add the names of newly assigned Soldiers if there are blank columns.

b. Under STATUS, record (*in pencil*) the date in the GO block if the Soldier demonstrated task proficiency to Soldier's manual standards. Keep this information current by always recording the most recent date on which the Soldier demonstrated task proficiency. Record the date in the NO GO block if the Soldier failed to demonstrate task proficiency to Soldier's manual standards. Soldiers who failed to perform the task should be retrained and re-evaluated until they can meet the standards. When that occurs, enter the date in the appropriate GO block and erase the previous entry from the NO GO block.

B-4. After the evaluation.

a. Read down each column (GO/NO GO) to determine the training status of that individual. This will give you a quick indication on which tasks a Soldier needs training.

b. Read across the rows for each task to determine the training status of all Soldiers. You can readily see on which tasks to focus training.

c. Line through the training status column of any Soldier who departs from the unit.

Figure B-1 is a sample of a completed DA Form 5165-R.

FIELD EXPEDIENT SQUAD BOOK For use of this form see AR 350-57. The proponent agency is DCSOPS.																	SHEET 1 OF 5	
USER APPLICATION	SOLDIER'S NAME																	
STP 11-31P14 MICROWAVE SYSTEMS OPERATOR-MAINTAINER	MOEHLMAN		COOPER		ESKEW		L. SMITH		PRUITT									
TASK NUMBER AND SHORT TITLE	STATUS		STATUS		STATUS		STATUS		STATUS		STATUS		STATUS		STATUS		STATUS	
	GO	NO GO	GO	NO GO	GO	NO GO	GO	NO GO	GO	NO GO	GO	NO GO	GO	NO GO	GO	NO GO	GO	NO GO
113-580-0053 TRBLSHT A TACLAN CABLE	4-2-03		3-2-03		3-2-03		3-2-03											
113-587-0074 SYS TRBLSHT RADIO AN/PRC-127	6-7-03		3-7-03		3-7-03		3-7-03											
113-587-1071 INSTALL MULTIPLEXER TD-1456/VRC	5-12-0		5-12-0		5-12-0		5-12-0											
113-587-2081 MAINTAIN SECURE AN/VRC-49 RETRANS	5-17-0		5-17-0		5-17-0		5-17-0											
113-587-2082 MAINTAIN SECURE AN/VRC-49 RETRANS																		
113-589-1009 INSTALL SECURE AN/VSC-7																		
113-587-2083 MAINTAIN SECURE TACSAT RETRANS																		
113-580-1034 INSTALL ATCCS CHS																		
113-580-0040 TRBLSHT MCS																		
113-580-3053 PERFORM SCHEDULED ULM ON MCS																		
113-580-0052 TRBLSHT SICPS CHS																		
113-580-0044 TRBLSHT PLGR																		

DA FORM 5165-R, SEP 85 (EDITION OF DEC 82 IS OBSOLETE)

Figure B-1. Sample DA Form 5165-R

GLOSSARY

Section I
Acronyms & Abbreviations

#	number
e.g.	for example
i.e.	that is
(C)	CONFIDENTIAL
(O)	FOR OFFICIAL USE ONLY
(S)	SECRET
(TS)	TOP SECRET
(U)	Unclassified
(V)	version
AAR	after action review
AC	alternating current/Active Component/assistant commandant
ACC	Army Correspondence Course/Automatic Color Control
ACCP	Army Correspondence Course Program
ack	acknowledge
ACP	Allied Communication Publication
ADC	air defense commander; area damage control
AF	audio frequency/Air Force
AFATDS	Advanced Field Artillery Tactical Data System
AGC	automatic gain control; assign gateway classmarks
AIT	Advanced Individual Training
AK	commercial cargo ship
ALE	automatic link establishment
ALM	Alarm
AM	amplitude modulation
AM/FM	amplitude modulation/frequency modulation

AME	antenna-mounted electronics
AMP/amp	amplifier; ampere(s)
AN	Annually (frequency code)
ANCD	automated net control device
ANCOC	Advanced Noncommissioned Officer Course
ANT	antenna
AOR	area of responsibility
app	appendix
AR	Army Regulation; Army Reserve
ARNG	Army National Guard
ASAS	All Source Analysis System
ASCII	America standard code for information interchange
ASIP	Advanced System Improvement Plan or Program
ATTN/attn	attention
AUTODIN	automatic digital network
aux	auxiliary
BA	biannually (frequency code)
BAT/BATT	battery
BCIS	Battlefield Combat Identification System
Bde	Brigade
BECS	Battlefield Electronic CEOI System
BER	bit error rate
BIT	built-in test
BITE	built-in test equipment
BM	bimonthly (frequency code)
BNCOC	Basic Noncommissioned Officer Course
BW	Biweekly (frequency code)
C	Celsius; change

C3	command, control, and communications
CADRG	Compressed Arc Digital Raster Graphic
CCU	communications control unit; camera control unit
CD	compact disk
CE	communications-electronics; common emitter
CFF	call for fire
chan	channel
CHAP/Chap	chapter
CIK	crypto/cryptographic ignition key
CM/cm	communications modem; center of mass
CMD	command; color monitor device
COML	commercial
COMM	communications
COMSEC	communications security
CONUS	Continental United States
CP	control processor; Command Post
CPU	central processing unit
CSM	Command Sergeant Major
CSMA	carrier-sensor multiple access
CSSCS	Combat Service Support Control System
CTC	Combat Training Centers
CTIL	commander's traced items list
CUM	Computer User's Manual
CW	continuous wave; clockwise
DA	Department of the Army; distribution amplifier
DA PAM	Department of the Army Pamphlet
DAMA	Demand Assigned Multiple Access
DC	direct current; District of Columbia; Dental Corps

DD	Department of Defense
DD Form	Department of Defense Form
DECRU	Downsize Enhanced Command Response Unit
DIV ID	division identifier
DMS	Defense Message System; degrees, minutes, seconds
DOS	disk operating system
DPS	data processing set
DS	Direct Support
DTD	data transfer device
DTED	digital terrain elevation data
DTG	date-time group; digital transmission group
DU	display unit
EA/ea	electronic attack; each
ECCM	electronic counter-counter measures
ECM	electronic countermeasures
ECU	environmental control unit
EEFI	essential elements of friendly information
EOM	end of message; end of mission
EP	electronic protection
EPLRS	Enhanced Position Location Reporting System
ESD	electrostatic discharge; electronic sensitive device
EW	electronic warfare
F	Fahrenheit; fail
FAAD	Forward Area Air Defense
FBCB2	Force XXI Battle Command Brigade and Below
FH	frequency hopping
FHMUX	frequency hopping multiplexer
FIPR	flash, immediate, priority, routine

FM	field manual; frequency modulation; file maintenance
FRAGO	fragmentary order
FREQ	frequency
G	green
GA	Georgia
GMT	Greenwich Mean Time
GPS	Global Positioning System
GR-RS	grid reference radio set
GTA	Graphic Training Aid
HF	high frequency
HI	high
HMMWV	high mobility multipurpose-wheeled vehicle
HQ	Headquarters
IAW	in accordance with
ICOM	integrated communications
ICT	intelligent communications terminal
IHFR	improved high frequency radio
Incl	Inclosure
IP	Internet protocol; implementation procedures
ISA	International Standardization Agreement; Industry Standard Architecture
JEWC	Joint Electronic Warfare Center
JINTACCS	Joint Interoperability of Tactical Command and Control Systems
JTF	joint task force
Kbd	keyboard
kHz	kilohertz
KSR	keyboard send/receive
KU	keyboard unit
KW/kw	kilowatt

L	left
LAN	local area network
LAT	latitude
LAT/LONG	latitude/longitude
LCN	logical channel number
LED	light emitting diode
LO	low
LOG/log	logistics
LOI	letter of instruction
LONG	longitude
LOS	line of sight
LRA	local reproduction authorized
LRBSDS	Long-Range Biological Standoff Detection System
LRF	laser range finder
LRU	lowest repairable unit
LSB	lower side band
M	medium; mandatory
MA	machine acknowledge
MAC	Maintenance Allocation Chart
MACOM	major Army command
MAX	maximum
MEDEVAC	medical evacuation
METL	mission essential task list
MFG	manufacturer
MGRS	Military Grid Reference System
MHz	megahertz
MICAD	multipurpose integrated chemical agent detector
MIJI	meaconing, intrusion, jamming, and interference

MO	Monthly (frequency code)
MOM	Manufacture's Operation Manual
MOOTW	military operations other than war
MOPP	mission-oriented protection posture
MOS	Military Occupational Specialty
MS-DOS	Microsoft-Disk Operating System
MSE	mobile subscriber equipment
MSG/Msg	message
MTO	message to observe
MTP	Mission Training Plan; MOS Training Plan
NA	not applicable
NATO	North Atlantic Treaty Organization
NAV	navigation
NAVAIDS	navigational aids
NBC	nuclear, biological, chemical
NCIM	Network Card Installation Manual
NCO	noncommissioned officer
NCOIC	noncommissioned officer in charge
NCS	net control station
NSA	National Security Agency
NSUM	Network Software User's Manual
NVIS	near vertical incidence skywave
O	Optional
OA	operator acknowledge
OG	operations group; optional group
OM	operation/operator's manual
OPCODE	operations code
OPFAC	operational facility

OPLAN	operation plan
OPM	Office of Personnel Management
OPORD	operation order
OPS	operational project stock; operations
OR	Ocean Region, operator response
OS	operating system
OTAR	over-the-air-rekey
OVM	operator vehicle maintenance; organizational vehicle maintenance
P	pass
PAM/Pam	power amplifier; port adapter module; pulse amplitude modulation; pamphlet
para	paragraph
PC	programmable controller; personal computer
PCC	printed circuit card
PCI	peripheral component interconnect
PLGR	Precision Lightweight Global Positioning System Receiver
PM	preventive maintenance; performance measure(s)
PMCS	preventive maintenance checks and services
PTT	push-to-talk
PU	power unit, processor unit
PUM	Printer User's Manual
PWR	power
QT	Quarterly (frequency code)
RC	Reserve Component; remote control; resistance-capacitance
RCVD	received
RECON	reconnaissance
RETRANS	retransmission station
RF	Reserve Forces; radio frequency

RFD	radio frequency direction
RHDDC	removable hard disk cartridge
RO	receive only
ROM	read only memory; Rough Order of Magnitude; Route of March
rpm	revolutions per minute
RPT	report
RS	radio set
RT	receiver-transmitter/radio transmitter
RTEP	real-time EPLRS program
RXMT	retransmit
S3	Operations and Training Officer (U.S. Army)
S-3	battalion or brigade operations staff officer (Army; Marine Corps battalion or regiment)
SA	situational awareness; semiannually (frequency code)
SALT	size, activity, location, and time
SATCOM	satellite communication(s)
SB	Supply Bulletin
SC	Signal Corps; single-channel; supply catalog; subcarrier; Service Code
SCRU	SINCGARS Remote Control Unit
SF	Standard Form
SI	Skill Identifier
SIG	Signal
SINCGARS	Single-Channel Ground and Airborne Radio System
SIP	System Improvement Program
SITREP	Situation Report
SL	skill level
SLDP	System Level Diagnostic Program
SM	soldier's manual
SOI	signal operation instructions

SOP	standard operating procedure
SPA	special psychological operations (PSYOP) assessment; submarine patrol area
SPKR	speaker
SQ ON	Squelch On
SRCU	SINCGARS Remote Control Unit
SRU	search and rescue unit
STP	soldier training publication
SUP/Suppl	Supplement
Sust	sustainment
TACSAT	tactical satellite
TADSS	Training Aids, Devices, Simulators, and Simulations
TB	technical bulletin
TC	technical coordinator; training circular; thermocouple
TEK	trunk encryption key
TFOM	time figure of merit
TG	trainer's guide
TI	Tactical Internet; test instrument; technical inspection
TM	technical manual
TOE	table(s) of organization and equipment
TRADOC	Training and Doctrine Command
TS	TOP SECRET; transmitter-to-studio
TTP	tactics, techniques, and procedures
TTY	teletypewriter
TX	Texas
UG	User's Guide
ULLS	Unit Level Logistics System
ULM	unit level maintenance

UNIT	trained in the unit (brevity code)
UPS	uninterruptible power supply
URN	unit requirement number
URO	user reader out
US	United States
USATSC	U.S. Army Training Support Command
USB	upper side band
USMTF	United States Message Text Format
UTM	Universal Traverse Mercator
UTO	unit task organization
VA	Virginia
VAA	vehicular amplifier adapter
VDC	volts direct current
VFTG	voice frequency telegraph terminal
VHF	very high frequency
Vkb	virtual keyboard
VOCO	verbal orders of commanding officer
Vol	Volume
VPF	Vector Product Format
Win	Windows
WK	Weekly (frequency code)
wpm	words per minute
WUM	Windows User's Manual
WWW	World Wide Web
XSTM	X-Windows System Tool Manager

Section II
Terms

additional skill identifier (ASI)
Identification of specialized skills that are closely related to and are in addition to those required by MOS or specialty skill identifier (SSI). Specialized skills identified by the ASI include operation and maintenance of specific weapons systems and equipment, administrative-type systems and subsystems, computer programming languages, procedures, installation management, analytic methods, animal handling techniques, and similar required skills that are too restricted in scope to comprise an MOS or SSI.

adjust
To maintain, within prescribed limits, by bringing into proper or exact position, or by setting the operating characteristics to the specified parameters.

alternating current (AC)
Electrical energy as supplied by normal wall outlets.

amplifier
A device which draws power from a source other than the input signal and which produces as an output an enlarged reproduction of the essential features of its input.

amplifier (amp)
A device or circuit that can increase the magnitude or power level of a time-variable signal without distorting its wave shape.

amplitude
The magnitude of variation in a changing quantity from its zero value.

amplitude modulation (AM)
The process of impressing information on a radio frequency signal by varying its amplitude.

analog
An analog signal that fluctuates exactly like the original stimulus.

Analog Signal
A signal used in communications lines that consist of a continuous electrical wave.

Antenna polarization
Physical orientation (vertical or horizontal) of a transmitting or receiving antenna.

Army Training and Evaluation Program (ARTEP)
The US Army's collective training program. The ARTEP establishes unit-training objectives critical to unit survival and performance in combat. They combine the training and the evaluation processes into one integrated function. The ARTEP is a training program and not a test. The sole purpose of external evaluation under this program is to diagnose unit requirements for future training.

automatic gain control
A type of circuit used to maintain the output volume of a receiver constant, regardless of variations in the signal strength applied to the receiver

bandwidth
Term used to define the frequency occupied by a signal and required for the effective transfer of information to be carried by that signal.

STP 11-25C13-SM-TG

baud
A variable unit of data transmission speed usually equal to one bit per second.

bias
(1) A voltage applied to a device (as a transistor controlled electrode) to establish a reference level for operation.
(2) A high frequency voltage combined with an audio signal to reduce distortion in tape recording.

bit (binary digit)
The smallest unit of information in a computer, a 1 or a 0. It can define two conditions (on or off).

bit per second (bps)
A measure of transmission speed to how many bits of binary data can be transmitted in one second.

BIT(S)
An element of a byte that can represent one of two values, on or off. There are 8 bits in a byte.

block diagram
A circuit diagram showing the general configuration of a piece of electronic equipment without showing the individual components. Lines or arrows may indicate the path of the signal or energy.

Call sign
Any combination of characters/numbers or pronounceable words that identifies a communication facility, command authority, activity, or unit.

circuit
A communications capability between two or more users, a user terminal and a switching facility, or between two switching facilities.

collective training
Training in units to prepare cohesive teams and units to accomplish their combined arms and services missions on the integrated battlefield.

common task
A critical task for which all soldiers at a given skill level are accountable, regardless of their MOS.

Cue
A word, situation, or other signal for action. An initiating cue is a signal to begin performing a task or task performance step. An internal cue is a signal to go from one element of a task to another. A terminating cue indicates task completion.

gain
Level of amplification for audio and video signals. "Riding the gain" is used in audio, meaning to keep the sound volume at a proper level.

individual training
Training which the officer, NCO, or soldier receives in the training base, units, on the job, or by self-study. This training prepares the individual to perform specified duties or tasks related to the assigned or next higher specialty code of MOS skill level and duty position.

Individual Training Evaluation Program (ITEP)
A program that requires commanders to routinely evaluate the soldier's ability to perform nonspecific tasks critical to the unit mission.

integration training
The completion of initial entry training in skill level 1 tasks for an individual newly arrived in a unit, but limited specifically to tasks associated with the mission, organization, and equipment of the unit to which the individual is assigned. It may be conducted by the unit using training materials supplied by TRADOC, by troop schools, or by in-service or contract mobile training teams. In all cases, the TRADOC school proponent supports this training.

Internet Controller (INC)
Part of the SINCGARS SIP program. The INC is mounted in the SINCGARS SIP vehicle amplifier adapter (VAA)

Liquid crystal diode (LCD)
Type of flat panel display screen that has liquid crystal deposited between two sheets of polarizing material. When an electrical current passes between crossing wires, the liquid crystals are aligned so light cannot shine through, producing an image on the screen.

Maintenance
The process of keeping equipment (or program) in working order.

MIJIFEEDER
A complete report of a MIJI incident

military occupational specialty (MOS)
The grouping of duty positions requiring similar qualifications and the performance of closely related duties.

Military Occupational Specialty (MOS) Code
A fixed number which indicates a given military occupational specialty. Also known as military occupational number and specification serial number.

MOS training plan (MTP)
The MTP is a guide for the conduct of individual training in units. The MTP is developed for each MOS/AOC and addresses all skill levels of an MOS/AOC and all duty positions. The MTP lists all MOS/AOC-specific and shared critical tasks for which the MOS/AOC is responsible. It will not include common tasks.

Radio Set Identifier (RSID)
The identifier that the EPLRS system uses to identify a specific radio in a network. The RS ID is unique for each radio and cannot be reused within the EPLRS community.

self-development
A planned, progressive, and sequential program followed by leaders to enhance and sustain their military competencies. Self-development consists of individual study, research, professional reading, practice, and self-assessment.

skill level (SL)
A number that denotes the level of qualification within the total MOS. Characters 0 through 5 in the position of the MOS code identify levels of qualification.

special qualifications identifiers (SQI)
(warrant officers and enlisted personnel only) an identification of skills in addition to those of an MOS to identify special requirements of certain positions and special qualifications of personnel who are capable of filling such positions. SQIs are authorized for use with any MOS unless otherwise specified.

training objective
A statement that describes the desired outcome of a training activity in the unit. It consists of the following three parts: task, condition(s), and standards.

train-up
The process of increasing the skills and knowledge of an individual to a higher skill level in the appropriate MOS. (It may involve certification.)

User
Any person or organization that needs or uses a terminal (attached to a computer).

STP 11-25C13-SM-TG

REFERENCES

Required Publications

Required publications are sources that users must read in order to understand or to comply with this publication.

Army Regulations

AR 380-5	Department of the Army Information Security Program (Item only produced in electronic media and included on EM 0001). 29 September 2000
AR 5-12	Army Management of the Electromagnetic Spectrum (this item is included on EM 0001). 1 October 1997
AR 750-1	Army Materiel Maintenance Policy. 18 August 2003

Department of the Army Forms

DA FORM 2028	Recommended Changes to Publications and Blank Forms. 1 February 1974
DA FORM 2404	Equipment Inspection and Maintenance Worksheet (Item only produced in electronic media and included on EM 0001). 1 April 1979
DA FORM 2407	Maintenance Request. 7 July 1994
DA FORM 2653-R	COMSEC Account – Daily Shift Inventory (LRA) (This item is included on EM 0001). 1 November 1974
DA FORM 5164-R	Hands-On Evaluation (LRA) (This item is included on EM 0001). 1 September 1985
DA FORM 5165-R	Field Expedient Squad Book. 1 September 1985
DA FORM 5986-E	Preventive Maintenance Schedule and Record (EGA). 1 March 1991
DA FORM 5988-E	Equipment Inspection Maintenance Worksheet (EGA). 1 March 1991

Department of Army Pamphlets

DA PAM 25-30	Consolidated Index of Army Publications and Blank Forms (Issued Quarterly) (No Printed Copies Exist) (Formerly DA PAM 310-1) (Item only produced in electronic media). 1 October 2004
DA PAM 25-7	Joint User Handbook for Message Text Formats (JUH-MTF). 1 July 1989
DA PAM 351-20	Army Correspondence Course Program Catalog. 1 October 1999
DA PAM 738-750	Functional Users Manual for the Army Maintenance Management System (TAMMS) (Item only produced in format). 1 August 1994

Department of Defense Forms

DD FORM 314	Preventive Maintenance Schedule and Record. 1 December 1953

Field Manuals

FM 101-5	Staff Organization and Operations. (Item also produced in electronic media.) 31 May 1997
FM 11-24	Signal Tactical Satellite Company. 30 September 1985
FM 11-32	Combat Net Radio Operations. 15 October 1990
FM 11-487-1	Installation Practices: HF Radio Communications Systems. 18 April 1991
FM 11-65	High Frequency Radio Communications. 31 October 1978
FM 24-1	Signal Support in the AirLand Battle. 15 October 1990

STP 11-25C13-SM-TG

FM 24-11	Tactical Satellite Communications (This item is included on EM 0205). 20 September 1990
FM 24-16	Communications-Electronics Operations, Orders, Records, and Reports. 7 April 1978
FM 24-18	Tactical Single-Channel Radio Communications Techniques (This item is included on EM 0205). 30 September 1987
FM 24-19	Radio Operator's Handbook. 24 May 1991
FM 24-22	Communications-Electronics Management System (CEMS). 30 June 1977
FM 24-33	Communications Techniques: Electronic Counter-Countermeasures (This item is included on EM 0205). 17 July 1990
FM 24-35	(O) Signal Operation Instructions "The SOI." 26 October 1990
FM 24-35-1	(O) Signal Supplemental Instructions. 2 October 1990
FM 24-7	Tactical Local Area Network (LAN) Management. 8 October 1999
FM 3-25.26	Map Reading and Land Navigation (This item is included on EM 0205). 20 July 2001

Other Product Types

ACP 124(D)	Communication Instructions Radiotelegraph Procedure. 1 October 1983.
ACP 125 US SUPPL-1	Communications Instructions Radiotelephone Procedures for Use by United States Ground Forces. 1 October 1985
ACP 125(E)	Communication Instructions Radiotelephone Procedures. 1 August 1987.
ACP 126(C)	Communication Instructions Teletypewriter (Teleprinter) Procedures. May 1989.
BFACS SUM	Battlefield Functional Area Control System (BFACS) Software User's Manual (SUM)
CDRL GO17 FBCB2 UMM	Contract Data Requirements List (CDRL) GO17 FBCB2 Unit Maintenance Manual (UMM)
CUM	Computer User's Manual (CUM)
DOS MANUAL	Disk Operating System (DOS) (or other system) Manual
EQUIPMENT TM(S)	Equipment Technical Manual(s)
MFG MANUALS	Manufacturer's manuals, issued with initial issue of equipment
MOM	Manufacture's/Monitor User's Manual (MOM)
NCIM	Network Card Installation Manual (NCIM)
NSUM	Network Software User's Manual (NSUM)
OM	Operator's Manuals (OM), KL-43 TRW Electronic Products Inc
PUB # 10515-0103-4100	AN/PRC-150(V)(C) Manpack Radio
PUM	Printer User's Manual (PUM)
SOI KTC 1400(*)	Numeral Cipher/Authentication System
SOI KTC 600(*)	Tactical Operations Code
SUM	Software User's Manuals (SUM)
UNIT OPLAN	Unit/Unit's Operation Plan (OPLAN)
UNIT OPORD	Unit/Unit's Operation Order (OPORD)
UNIT SOI	Unit/Unit's Signal Operation Instructions (SOI)
UNIT SOP	Unit/Unit's Standing Operating Procedure (SOP)
UNIT SUPPLY UPDATE	Unit Supply Update
UNIT TACTICAL SOP	Unit's Tactical Standing Operating Procedures (SOP)
VDC-300	Operator's Manual
VIASAT	ViaSat E-mail User's Guide

STP 11-25C13-SM-TG

| WUM | WINDOWS User's Manual (WUM) |

Soldier Training Publications

| STP 21-1-SMCT | Soldier's Manual of Common Tasks Skill Level 1. (Item also produced in electronic media.) 31 August 2003 |
| STP 21-24-SMCT | Soldier's Manual of Common Tasks (SMCT) Skill Levels 2-4. (Item also produced in electronic media.) 31 August 2003 |

Supply Bulletins

| SB 11-131-1 | Vehicular Sets and Authorized Installations (Volume I) (This item is included on EM 0168). 1 March 1991 |
| SB 11-131-2 | Vehicular Radio Sets and Authorized Installations Volume II (SINCGARS, FHMUX, and EPLRS) (This item is included on EM 0144). 1 July 1998 |

Technical Bulletins

TB 11-5820-890-12	Operator and Unit Maintenance for AN/CYZ-10 Automated Net Control Device (ANCD) with the Single Channel Ground and Airborne Radio Systems (SINCGARS) (Item also produced in electronic media and included on EM 0071). 1 April 1993
TB 11-5825-283-23	Not found. Unit Operation and Maintenance for Enhanced Position Location Reporting System Grid Reference Unit
TB 380-41	Security: Procedures for Safeguarding, Accounting, and Supply Control of COMSEC Material (This item is included on EM 0248). 3 July 2003
TB 43-0129	Safety Requirements for Use of Antenna and Mast Equipment. (Item also produced in electronic media and included on EM 0161.) 15 June 1986

Technical Manuals

DTM 11-5810-394-14&P	Operator's Unit, Direct Support, and Special Repair Activity Maintenance Manual (Including Repair Parts and Special Tools) for the Automated Net Control Device (ANCD) AN/CYZ-10. 14 January 1994
TM 11-5825-295-10	Operator's Manual for Net Control Station AN/TSQ-158(V)4. 15 April 2001
TM 11-5825-295-23&P	Unit and Direct Support Maintenance Manual for Net Control Station AN/TSQ-158(V)4. 1 April 2001
DTM 11-7010-326-10	Operator's Manual Force XXI Battle Command Brigade and Below (FBCB2). 22 December 2000
DTM 11-7010-326-20&P	Unit Level Maintenance Manual for Force XXI Battle Command Brigade and Below (FBCB2). 22 December 2000
TM 10-5411-222-14	Operator's, Unit, Direct Support, and General Support Maintenance Manual for Shelter, Nonexpendable Integrated, S-787/G (This item is included on EM 0153). 1 April 2003
TM 11-2300-475-13&P-1	Operator's, Organizational, & DS Maint Manual (Incl Repr Parts & Special Tools List) for Install Kit, Elect Eq MK-2462/GRC-193A in Truck Cargo, 1 1/4 Ton, 4x4, M882 or M1008A1 (CUCV)....Radio Set AN/GRC193A (Reprinted w/Basic Incl C1). 1 September 1987
TM 11-2300-476-14&P	Operator's, Unit, Direct Support and General Support Maintenance Manual (Including Repair Parts and Special Tools List) for Installation Kits, Electronic Equipment: MK-2442/GRC-213 for Armored Personnel Carrier-M113A1; MK-2443/GRC-213 ..... 1 January 1987
TM 11-5805-201-12	Operator's and Unit Maintenance Manual for Telephone Sets, TA-312/PT and TA-312A/PT (This item is included on EM 0059). 1 August 1990

TM 11-5805-780-15	Digital Nonsecure Voice Terminal (DNVT) with Digital Data Port TA-1042A/U. 15 May 1992
TM 11-5815-334-10	Operator's Manual for Radio Teletypewriter Sets, AN/GRC-122, AN/GRC-122A, AN/GRC-122B, AN/GRC-122C, AN/GRC-122D, AN/GRC-122E, AN/GRC-142, AN/GRC-142A, AN/GRC-142B, AN/GRC-142C, AN/GRC-142D, and AN/GRC-142E...(Reprinted w/Basic Incl C1-2). 5 March 1985
TM 11-5815-602-10	Operator's Manual for Terminal Communications AN/UGC-74A(V)3. 23 September 1983
TM 11-5815-602-10-1	Operator's Manual for Terminal, Communications AN/UGC-74B(V)3, Terminal, Communications AN/UGC-74C(V)3. 1 March 1987
TM 11-5815-602-24	Organizational, Direct Support and General Support Maintenance Manual for Terminal, Communications AN/UGC-74A(V)3. 8 January 1984
TM 11-5820-1130-12&P	Operator's and Unit Maintenance Manual (Including Repair Parts and Special Tools List) for Radio Set AN/PSC-5. 1 June 2000
TM 11-5820-256-10	Operator's Manual: Radio Set, AN/GRC-26D. 19 June 1969.
TM 11-5820-467-15	Operator's, Organizational, Direct Support, General Support, and Depot Maintenance Manual for Antenna Group, AN/GRA-50 (Reprinted w/Basic Incl C3-7). 19 July 1961
TM 11-5820-477-12	Operator's and Organizational Maintenance Manual Radio Set Control Groups AN/GRA-39, AN/GRA-39A, and AN/GRA-39B (Reprinted w/Basic Incl C1-3). 10 July 1975
TM 11-5820-520-10	Operator's Manual for Radio Sets, AN/GRC-106 and AN/GRC-106A (Reprinted w/Basic Incl C1). 28 May 1984
TM 11-5820-890-10-1	Operator's Manual for SINCGARS Ground Combat Net Radio, ICOM Manpack Radio AN/PRC-119A, Short Range Vehicular Radio AN/VRC-87A, Short Range Vehicular Radio With Single Radio Mount AN/VRC-87C, Short Range...Range Vehicular Radio AN/VRC-92A. 1 September 1992
TM 11-5820-890-10-3	Operator's Manual for SINCGARS Ground Combat Net Radio, Non-ICOM Manpack Radio AN/PRC-119, Short Range Vehicular Radio AN/VRC-87, Short Range Vehicular Radio (with Single Radio Mount) AN/VRC-87D....Long Range Vehicular Radio AN/VRC-92. 1 September 1992
TM 11-5820-890-10-8	Operator's Manual for SINCGARS Ground Combat Net Radio, ICOM Manpack Radio, AN/PRC-119A, Short Range Vehicular Radio AN/VRC-87A, Short Range, Multiplexer (FHMUX) (Item also produced in electronic media and included on EM 0071). 1 December 1998
TM 11-5820-890-20-1	Unit Maintenance Manual Ground ICOM Radio Sets AN/PRC-119A, AN/PRC-119D, AN/PRC-119F, AN/VRC-87A, AN/VRC-87D, AN/VRC-87F, AN/VRC-88A, AN/VRC-88D, AN/VRC-88F, AN/VRC-89A, AN/VRC-89D, AN/VRC-89F, AN/VRC-90A, AN/VRC-90D, AN/VRC-90F, ... 1 December 1998
TM 11-5820-890-20-2	Unit Maintenance Manual for Ground ICOM Radio Sets AN/PRC-119A, AN/VRC-119D, AN/VRC-87A, AN/VRC-87C, AN/VRC-87D, AN/VRC-88A, AN/VRC-88D, AN/VRC-89A, AN/VRC-89D, AN/VRC-90A, AN/VRC-90D, AN/VRC-91A, AN/VRC-91D, AN/VRC-92A, AN/VRC-92D (With... 1 July 2000

STP 11-25C13-SM-TG

TM 11-5820-890-20-3	Unit Level Maintenance Handbook, SINCGARS ICOM Ground Radios, Ground ICOM Radio Sets, AN/PRC-119A, AN/VRC-87A, AN/VRC-88A, AN/VRC-89A, AN/VRC-90A, AN/VRC-91A, AN/VRC-92A. 28 February 1995
TM 11-5820-919-12	Operator's and Organizational Maintenance Manual for Radio Set AN/PRC-104(A). 15 January 1986
TM 11-5820-923-12	Operator's and Organizational Maintenance Manual for Radio Set AN/GRC-213 (Reprinted w/Basic Incl C1-2). 14 February 1986
TM 11-5820-924-13	Operator's, Organizational and Direct Support Maintenance Manual for Radio Set, AN/GRC-193A (Reprinted w/Basic Incl C1). 14 February 1986
TM 11-5825-283-10	Operator's Manual for Manpack Radio Sets AN/ASQ-177C(V)4; AN/PSQ-6C; AN/VSQ-2C(V)1; AN/VSQ-2C(V)2; AN/VSQ-2C(V)4; Grid Reference Radio Set AN/GRC-229C; Downsized Enhanced Command Response Unit RT-1718/TSQ-158A. 15 August 2000
TM 11-5825-283-10-1	Operator's Manual for Radio Sets AN/ASQ-177(V)3, AN/ASQ-177(V)4, AN/PSQ-6, AN/VSQ-2(V)1, AN/VSQ-2(V)2, AN/VSQ-2(V)3, AN/VSQ-2(V)4, and Grid Reference Radio Set AN/GRC-229 as Part of Enhanced Position Location Reporting System. 1 August 2000
TM 11-5825-283-23	Unit and DS Maintenance Manual for Radio Sets AN/ASQ-177(V)3, AN/ASQ-177(V)4, AN/PSQ-6, AN/VSQ-2(V)1, AN/VSQ-2(V)2, AN/VSQ-2(V)3, AN/VSQ-2(V)4, and Grid Reference Radio Set AN/GRC-229 as Part of Enhanced Position Location Reporting.... 1 April 1994
TM 11-5825-291-13	Operations and Maintenance Manual for Satellite Signals Navigation Sets AN/PSN-11 and AN/PSN-11(V)1. 1 April 2001
TM 11-5825-295-10	Operator's Manual for Net Control Station AN/TSQ-158(V)4, Part of Enhanced Position Location Reporting System. 1 April 2001
TM 11-5825-299-10	Operator's Manual for Airborne Radio Set AN/ASQ-177D(V)4; Ground Radio Sets AN/PSQ-6D, AN/VSQ-2D(V)1, AN/VSQ-2D(V)2, AN/VSQ-2D(V)4 and Grid Reference Radio Set AN/GRC-229D. 31 July 2003
TM 11-5895-1180-20	Organizational Maintenance Manual for Radio Set AN/PSC-3 (Reprinted w/Basic Incl C1-2). 15 February 1988
TM 11-5895-1181-20	Organizational Maintenance Manual Radio Set AN/VSC-7 (Reprinted w/Basic Incl C1). 15 February 1988
TM 11-5895-1348-13&P	Operator's, Unit and Direct Support Maintenance Manual (Including Repair Parts and Special Tools List) for Tactical Computer Processor AN/UYQ-43(V)1. 15 October 1993
TM 11-5895-1546-12&P	Operator's and Unit Maintenance Manual Including Repair Parts and Special Tools List for Communications Modem, TCIM Assembly MD-1298/U, P/N 50766 TCIM 1 Assembly, P/N 50787; TCIM 2 Assembly, P/N 52605. 1 June 2000
TM 11-5985-357-13	Operator's, Organizational, and Direct Support Maintenance Manual for Antenna Group OE-254/GRC. 1 February 1991
TM 11-5985-426-12&P	Operator's and Unit Maintenance Manual Including Repair Parts and Special Tools List for Mast AB-1386/U; Electrical Equipment Mounting Bases MT-6967/G and MT-6968/G (Reprinted w/Basic Incl C1-4). 1 July 1996
TM 11-7010-326-10	Operator's Manual Force XXI Battle Command Brigade and Below (FBCB2). December 2000 (Draft)
TM 38-750	The Army Maintenance Management System (TAMMS). May 1968.
TM 5-6115-271-14	Operator's, Organizational, Direct Support and General Support Maintenance Manual: Generator Set, 3 KW, 28V DC (Less Engine) DOD Model MEP-016A, 60 HZ, Model MEP-016C 60 HZ, Model MEP-021A

	400 HZ, Model MEP-021C 400 HZ, Model MEP-026A DC HZ, Model MEP-026C 28 V DC (Reprinted w/Basic Incl C1-12) (This item is included on EM 0086). 31 March 1993
TM 5-6115-275-14	Operator's, Organizational, Intermediate (Field) (Direct Support and General Support) and Depot Maintenance Manual Generator Set, Gasoline Engine Driven, Skid Mounted, Tubular Frame, 10 KW, AC, 120/208V 3 Phase and 120/240V Single Phase, Less Engine: DOD Models MEP-018A, 60 HZ and MEP-023A, 400 HZ (Reprinted w/Basic Incl C1-9) (This item is included on EM 0086). 16 June 1977
TM 5-6115-332-14	Operator, Organizational, Intermediate (Field), Direct Support, General Support, and Depot Level Maintenance Manual: Generator Set, Tactical, Gasoline Engine: Air Cooled, 5 KW, AC, 120/240 V, Single Phase, 120/208 V, 3 Phase, Skid Mounted, Tubular Frame (Less Engine) (Military Design DOD Model MEP-017A), Utility, 60 HHZ and (DOD Model MEP-022A), Utility, 400 HZ (Reprinted w/Basic Incl C1-10) (This item is included on EM 0086). 9 December 1977
TM 5-6115-365-15	Operator's, Organizational, Direct Support, General Support and Depot Maintenance Manual Including Repair Parts and Special Tools List for Generator Sets, Gasoline and Diesel Engine Driven, Trailer Mounted, PU-236A/G, PU-236/G, PU-236B/G, P.... 11 May 1966
TM 5-6115-615-12	Operator and Organizational Maintenance Manual: Generator Set, Diesel Engine Driven, Tactical, Skid Mounted 3 KW, 3 Phase 120/208 and Single Phase 120/240 Volts AC and 28 Volts DC DOD Model MEP-016B; Class Utility; Mode 60 HZ; DOD Model..... 31 July 1987
TM 5-6115-640-14&P	Operator, Unit, Direct Support and General Support Maintenance Manual (Including Repair Parts and Special Tools List) for Power Plants AN/MJQ-32, AN/MJQ-33, MEP-701A 3 KW 60 HZ Acoustic Suppression Kit Generator Sets M116A2 2-Wheel..... 28 July 1989
TM 9-2330-202-14&P	Operator's, Unit, Direct Support, and General Support Maintenance Manual (Incl Repair Parts & Special Tools List) for Trailer: Cargo: 3/4 Ton, 2-Wheel M101A2, M101A3 Trailer: Chassis, 3/4 Ton, 2-Wheel M116A2....12 May 1997
TM 9-6115-641-10	Operator's Manual for Generator Set Skid Mounted, Tactical Quiet 5 KW, 60 and 400 HZ MEP-802A (60 HZ), MEP-812A (400 HZ) (Reprinted w/Basic Incl C1-2) (This item is included on EM 0086). 30 December 1992
TM 9-6115-642-10	Operator's Manual for Generator Set Skid Mounted, Tactical Quiet 10 KW, 60 and 400 HZ MEP-803A (60 HZ) MEP-813A (400 HZ) (Reprinted w/Basic Incl C1-2) (This item is included on EM 0086). 30 December 1992
TM 9-6115-659-13&P	Operator, Unit, and Direct Support Maintenance Manual (Including Repair Parts and Special Tools List) for Power Unit, Diesel Engine Driven, 1 Ton Trailer Mounted, 5 KW, 60 HZ, PU-797 Power Unit, Diesel Engine Driven, High Mobility Trailer Mounted, 5 KW, 60 HZ, PU-797A Power Plant....(This item is included on EM 0086). 15 October 1993
TM 9-6115-660-13&P	Operator, Unit, and Direct Support Maintenance Manual (Including Repair Parts and Special Tools List) For Power Unit, Diesel Engine Driven, 1 Ton Trailer Mounted, 10 KW, 60 HZ, PU-789,..... (Reprinted w/Basic Incl C1). 15 October 1993

Training Circulars

TC 24-21	Tactical Multichannel Radio Communications Techniques. (Item also produced in electronic media.) 3 October 1988

Training Support Packages

TSP 113-587-1067   Install Secure Single-Channel Ground and Airborne Radio System (SINCGARS) ICOM With or Without the AN/VIC-1. 1 March 1999

Related Publications

Related publications are sources of additional information. They are not required in order to understand this publication.

Army Correspondence Course Program Subcourses

SS 0018   Install Generator Set 5 KW or PU-620.
SS 0456   Communications Security (COMSEC).
SS 0733   Install Radio Teletypewriter Set AN/GRC-142(*) or AN/GRC-122(*).
SS 9740   Radio Set AN/GRC-106 (SOJT).
SS 9850   Troubleshoot Radio Teletypewriter Set AN/GRC-142(*) or AN/GRC-122(*).

Army Regulations

AR 25-11   Record Communications and the Privacy Communications System (Item only produced in electronic media and included on EM 0001). 4 September 1990
AR 380-5   Department of the Army Information Security Program (Item only produced in electronic media and included on EM 0001). 29 September 2000

Department of Army Forms

DA FORM 2062   Hand Receipt/Annex Number. 1 January 1982

Department of Army Pamphlets

DA PAM 738-750   Functional Users Manual for the Army Maintenance Management System (TAMMS) (Item only produced in format). 1 August 1994

Field Manuals

FM 11-32   Combat Net Radio Operations. 15 October 1990
FM 11-487-1   Installation Practices: HF Radio Communications Systems. 18 April 1991
FM 24-18   Tactical Single-Channel Radio Communications Techniques (This item is included on EM 0205). 30 September 1987
FM 24-19   Radio Operator's Handbook. 24 May 1991
FM 24-22   Communications-Electronics Management System (CEMS). 30 June 1977
FM 24-35-1   (O) Signal Supplemental Instructions. 2 October 1990
FM 3-25.26   Map Reading and Land Navigation (This item is included on EM 0205). 20 July 2001

Graphic Training Aids

GTA 11-3-20   Installation of Antenna Group OE-254/GRC

STP 11-25C13-SM-TG

Other Product Types

CUM	Computer User's Manual (CUM)
EQUIPMENT TM(S)	Equipment Technical Manual(s)
GPS-DC MK III OSM	Operating and Service Manual (OSM) for the GPS-DC MK III
MOM	Manufacture's/Monitor User's Manual (MOM)
PUM	Printer User's Manual (PUM)
WUM	WINDOWS User's Manual (WUM)

Supply Bulletins

SB 11-131-1	Vehicular Sets and Authorized Installations (Volume I) (This item is included on EM 0168). 1 March 1991

Standard Forms

SF 153 FORM	COMSEC Material Report. 18 March 1996
SF 700 FORM	Security Container Information. August 1985
SF 702 FORM	Security Container Check Sheet. August 1985

Technical Bulletins

TB 11-5820-890-12	Operator and Unit Maintenance for AN/CYZ-10 Automated Net Control Device (ANCD) with the Single Channel Ground and Airborne Radio Systems (SINCGARS) (Item also produced in electronic media and included on EM 0071). 1 April 1993

Technical Manuals

(O)TM 11-5810-394-14&P	Operator's Unit, Direct Support, and Special Repair Activity Maintenance Manual (Including Repair Parts and Special Tools) for the Automated Net Control Device (ANCD) AN/CYZ-10. 14 January 1995
DTM 11-7010-326-10	Operator's Manual Force XXI Battle Command Brigade and Below (FBCB2). 22 December 2000
TM 11-5815-334-10	Operator's Manual for Radio Teletypewriter Sets, AN/GRC-122, AN/GRC-122A, AN/GRC-122B, AN/GRC-122C, AN/GRC-122D, AN/GRC-122E, AN/GRC-142, AN/GRC-142A, AN/GRC-142B, AN/GRC-142C, AN/GRC-142D, and AN/GRC-142E...(Reprinted w/Basic Incl C1-2). 5 March 1985
TM 11-5815-602-10	Operator's Manual for Terminal Communications AN/UGC-74A(V)3 (This item is included on EM 0080). 23 September 1983
TM 11-5815-602-10-1	Operator's Manual for Terminal, Communications AN/UGC-74B(V)3, Terminal, Communications AN/UGC-74C(V)3 (This item is included on EM 0080). 1 March 1987
TM 11-5815-602-24	Organizational, Direct Support and General Support Maintenance Manual for Terminal, Communications AN/UGC-74A(V)3. 8 January 1984
TM 11-5820-1130-12&P	Operator's and Unit Maintenance Manual (Including Repair Parts and Special Tools List) for Radio Set AN/PSC-5 (This item is included on EM 0169). 1 June 2000
TM 11-5820-477-12	Operator's and Organizational Maintenance Manual Radio Set Control Groups AN/GRA-39, AN/GRA-39A, and AN/GRA-39B (Reprinted w/Basic Incl C1-3) (This item is included on EM 0079). 10 July 1975

STP 11-25C13-SM-TG

TM 11-5820-520-10	Operator's Manual for Radio Sets, AN/GRC-106 and AN/GRC-106A (Reprinted w/Basic Incl C1). 28 May 1984
TM 11-5820-890-10-1	Operator's Manual for SINCGARS Ground Combat Net Radio, ICOM Manpack Radio AN/PRC-119A, Short Range Vehicular Radio AN/VRC-87A, Short Range Vehicular Radio With Single Radio Mount AN/VRC-87C, Short Range...Range Vehicular Radio AN/VRC-92A. 1 September 1992
TM 11-5820-890-10-8	Operator's Manual for SINCGARS Ground Combat Net Radio, ICOM Manpack Radio, AN/PRC-119A, Short Range Vehicular Radio AN/VRC-87A, Short Range...... Multiplexer (FHMUX) (Item also produced in electronic media and included on EM 0071). 1 December 1998
TM 11-5820-919-12	Operator's and Organizational Maintenance Manual for Radio Set AN/PRC-104(A) (This item is included on EM 0140). 15 January 1986
TM 11-5820-923-12	Operator's and Organizational Maintenance Manual for Radio Set AN/GRC-213 (Reprinted w/Basic Incl C1-2) (This item is included on EM 0140). 14 February 1986
TM 11-5820-924-13	Operator's, Organizational and Direct Support Maintenance Manual for Radio Set, AN/GRC-193A (Reprinted w/Basic Incl C1) (This item is included on EM 0140). 14 February 1986
TM 11-5825-283-10	Operator's Manual for Manpack Radio Sets AN/ASQ-177C(V)4; AN/PSQ-6C; AN/VSQ-2C(V)1; AN/VSQ-2C(V)2; AN/VSQ-2C(V)4; Grid Reference Radio Set AN/GRC-229C; Downsized Enhanced Command Response Unit RT-1718/TSQ-158A. 15 August 2000
TM 11-5825-291-13	Operations and Maintenance Manual for Satellite Signals Navigation Sets AN/PSN-11 and AN/PSN-11(V)1. 1 April 2001
TM 11-5895-1499-12	Operator's and Unit Maintenance Manual for Communications Central AN/TSC-122 (Reprinted w/ Basic Inclosure C1). 15 September 1991
TM 11-5985-379-14&P	Operator's Organizational, Direct Support, and General Support Maintenance Manual (Including Repair Parts and Special Tools List) Antenna AS-2259/GRA. 14 February 1986
TM 11-6625-3015-14	Operator's, Organizational, Direct Support and General Support Maintenance Manual for Radio Test Set, AN/PRM-34 (Reprinted w/Basic Incl C1-4). 14 October 1983
TM 11-6625-3052-14	Operator's, Unit, Direct Support and General Support Maintenance Manual for Digital Multimeter AN/PSM-45 (Reprinted w/Basic Incl C1-2). 10 January 1984
TM 11-6625-3199-14	Operator's, Unit, Intermediate Direct Support, and General Support Maintenance Manual for Digital Multimeter AN/PSM-45A (Reprinted w/Basic Incl C1). 15 December 1988
TM 38-750	The Army Maintenance Management System (TAMMS). May 1968.
TM 5-4120-379-14	Operator's, Organizational, Direct Support and General Support Maintenance Manual for Air Conditioner, Horizontal Compact; 18,000 BTU/HR, 208V, 3 Phase; 50/60 Hz. 28 November 1983 (C1-4)
TM 5-4120-383-14	Operator's Organizational, Direct Support and General Support Maintenance Manual for Air Conditioner, Compact; 9,000 BTU/HR, 208 V, 3 Phase; 50/60 HZ, Model F9000H-3S (Reprinted with Basic Inclosure C1). 9 May 1985
TM 5-4120-384-14	Operator's, Organizational, Direct Support and General Support Maintenance Manual Air Conditioner, Horizontal, Compact; 18,000 BTU/HR, 208V, 3 Phase, 50/60 Hz, Model F18H-3S and 230V, Single Phase, 60 Hz, Mdl F18H-1S... (Reprinted w/Basic Incl C1-4). 27 May 1985

TM 5-6115-271-14	Operator's, Organizational, Direct Support and General Support Maintenance Manual: Generator Set, 3 KW, 28V DC (Less Engine) DOD Model MEP-016A, 60 HZ, Model MEP-016C 60 HZ, Model MEP-021A 400 HZ, Model MEP-021C 400 HZ, Model MEP-026A DC HZ, Model MEP-026C 28 V DC (Reprinted w/Basic Incl C1-12) (This item is included on EM 0086). 31 March 1993
TM 5-6115-275-14	Operator's, Organizational, Intermediate (Field) (Direct Support and General Support) and Depot Maintenance Manual Generator Set, Gasoline Engine Driven, Skid Mounted, Tubular Frame, 10 KW, AC, 120/208V 3 Phase and 120/240V Single Phase, Less Engine: DOD Models MEP-018A, 60 HZ and MEP-023A, 400 HZ (Reprinted w/Basic Incl C1-9) (This item is included on EM 0086). 16 June 1977
TM 5-6115-365-15	Operator's, Organizational, Direct Support, General Support and Depot Maintenance Manual Including Repair Parts and Special Tools List for Generator Sets, Gasoline and Diesel Engine Driven, Trailer Mounted, PU-236A/G, PU-236/G, PU-236B/G, P....(Reprinted w//Basic Incl C1-11) (This item is included on EM 0086) (NG) (AR) (IL). 11 May 1966

Training Circulars

TC 24-21	Tactical Multichannel Radio Communications Techniques. (Item also produced in electronic media.) 3 October 1988

Training Support Packages

TSP 113-587-1067	Install Secure Single-Channel Ground and Airborne Radio System (SINCGARS) ICOM With or Without the AN/VIC-1. 1 March 1999

STP 11-25C13-SM-TG
16 February 2005

By Order of the Secretary of the Army:

PETER J. SCHOOMAKER
General, United States Army
Chief of Staff

Official:

SANDRA R. RILEY
Administrative Assistant to the
Secretary of the Army
0502601

DISTRIBUTION:

Active Army, Army National Guard, and U.S. Army Reserve: To be distributed in accordance with the initial distribution number 112066, requirements for STP 11-31C13-SM-TG.

This page intentionally left blank.

PIN: 082167-000

CPSIA information can be obtained at www.ICGtesting.com
Printed in the USA
BVOW09s2113090315

390929BV00008BA/91/P